Springer Geology

Series Editors

The book series Springer Geology comprises a broad portfolio of scientific books, aiming at researchers, students, and everyone interested in geology. The series includes peer-reviewed monographs, edited volumes, textbooks, and conference proceedings. It covers the entire research area of geology including, but not limited to, economic geology, mineral resources, historical geology, quantitative geology, structural geology, geomorphology, paleontology, and sedimentology.

Michael George Petterson

Himalayan Thick-Skin Basement Deformation of the Ladakh Batholith, Leh-Ladakh Region, NW India

Springer

Michael George Petterson
School of Science
Auckland University of Technology
Auckland, New Zealand

ISSN 2197-9545 ISSN 2197-9553 (electronic)
Springer Geology
ISBN 978-3-031-31565-7 ISBN 978-3-031-31566-4 (eBook)
https://doi.org/10.1007/978-3-031-31566-4

This Springer imprint is published by the registered company Springer Nature Switzerland AG
The registered company address is: Gewerbestrasse 11, 6330 Cham, Switzerland

This book is dedicated to my partner, Lanka Nanayakkara, who accompanied me with great patience and spirit, in all our Ladakh geological journeys, and without whose constant support and encouragement, this book would not have been written.

Preface

This book centres on the capital town of Leh within the Union Territory of Ladakh, NW India and extends many kilometres NW, SE and south of Leh itself into remote valleys, plains and mountains.

The book is written in an attempt to answer a number of geological research questions including: 'Is it possible to map the Ladakh batholith in such a way that it adds value to the already-existent knowledge base?'. Batholiths are complex and heterogeneous. Many geologists prefer to work in less complicated terranes or can dismiss batholiths as purely 'a lot of boring granite'. Researchers tend to take samples from here and there in a rather piecemeal fashion, dating rocks, measuring isotope ratios and chemical compositions and determining geophysical parameters. These data have greatly enhanced our understanding of the age span, origin, petrology, uplift history, previous plate tectonic geography and probable geotectonic origin of the batholith and surrounding terranes.

Few researchers, however, spend significant periods of time mapping and measuring the actual lithological variety of batholiths, or recording the many elements of observable magmatic, metamorphic and deformational structures systematically across a significant geographical area. This study presents a more detailed set of geoscientific field observations, and geological/structural maps and data, of the Ladakh batholith, than has occurred hitherto. The author believes that significant value has been added to the geological knowledge base and geological maps and hopes these can be used as a basis for future research in a number of multidisciplinary areas, as well as deepening understanding.

Perhaps the deepest research question centres upon 'Can we determine the deformation processes that have left their mark within the batholith?'. Unlike bedded cover sedimentary rocks that (relatively speaking) clearly deform into contiguous folds and faults that are directly observable, measurable, recordable and photographed, basement rocks are a different 'breed' altogether. The crystalline, heterogeneous, inconsistent nature of plutonic batholithic basement and the lack of pre-existing contiguous planar structures (such as bedding in sedimentary rocks) create an extra layer of complexity in understanding deformation processes and their effects. The Leh-Ladakh region lends itself to unravelling this more cryptic science, partly because of the high degree of rock exposure, and the close association between geomorphology and structure. This book reveals, through a field-geology and follow-up image analysis-based approach, that the batholith has been highly to significantly deformed and impacted by both brittle and ductile deformation. This deformation has resulted in the formation of a series of fractured crustal blocks that have moved vertically and laterally, interacting with one another at Klippen boundaries.

The book sets the research results within the context of previous research and a literature review. It is highly illustrated and presents a transparent approach for readers to decide for themselves and perhaps offer alternate explanations for the evidence. As a scientist, I have presented, to the best of my ability, ideas, hypotheses and explanations, for the data and landscape analysis, but also know that science and ideas are never static, and my models and ideas will be tested through rigorous reviews, research and reviews that will occur in the future.

Auckland, New Zealand Michael George Petterson

Acknowledgements

I acknowledge with gratitude, time and finances provided to me from my employers, the Auckland University of Technology, that enabled this work to be undertaken and written-up, particularly Hannah Buckley and Kay Vopel.

I deeply thank my partner Lanka Nanayakkara, who was with me in the field, and forever encouraged the writing of this book: without her it would not have been written.

I acknowledge and thank the numerous people of Ladakh who helped and assisted myself and my partner during our time in the field. I would particularly like to thank Rigzen Norbu who drove and guided us so professionally through many valleys and remote terrain. To T. A. Angmo for her kind hospitality and that of her family, and to Rahul who explained so much of Ladakhi culture to me. To the management and staff of the Blue Stone and Grand Dragon hotels in Leh, the latter of which is brave enough to stay open through the cold winter seasons, during which the work was undertaken. I also thank the people of Leh and Ladakh who were so helpful to us during our time there. Thank you to Graham Hinchliffe for all your great advice re all things technical GIS and imagery and to Profs. Brian Windley, M. Qasim Jan and M. Asif Khan who were a tremendous inspiration to me as a Ph.D. student and who introduced me to the beauty and geological wonders of the NW Himalayas.

Professor Petterson has published over 200 papers, geological maps, book chapters and popular articles. He has worked as a guest editor for the *Geoscience Journal*, held senior leadership positions with the British Geological Survey, Auckland University of Technology and the Secretariat of the Pacific Community, and residential or long-term professional positions in Solomon Islands, Guyana, Afghanistan and Fiji.

Introduction

The Himalaya, Trans-Himalaya, Karakoram and Hindu Kush mountains, together with the upraised plateau of Tibet, constitute the highest topographic orogenic belt on Earth. Created by the collision between two mega-plates (India and Eurasia) and a number of minor plates from c. 50–55 million years ago, they constitute a continental geographical barrier of planetary-impressive proportions.

To the south is the Indian subcontinent and the modern countries of Afghanistan, Pakistan, India, Nepal, Bhutan, Bangladesh and Myanmar. To the north is Tibet, Xinjiang, Mongolia and Siberia. The mountain chain continues westwards and eastwards, with the historical Silk Road using the passes and valleys of the Greater Himalaya for human transport and communications between civilizations across the centuries from Europe, Central Asia, India and China.

A careful study of the form of the Himalayas reveals a curious northward indentation, as far northwards as the main Himalayan mountains attain, in north Pakistan and NW India. Here the Himalaya, Karakoram and Hindu Kush mountain ranges meet. The formidable Nanga Parbat mountain forms the apex of the Himalayan mountains in this region, created by a remarkable tectonic process that has rapidly uplifted great depths of geological basement to the surface, producing a living, dynamic and active, tectonic landscape.

Within this NW Himalayan landscape is the ancient Buddhist kingdom of Ladakh, now a Unitary Territory of India, previously a part of Jammu and Kashmir. This stunningly beautiful high mountain desert landscape, with elevated valleys adjacent to the mighty River Indus, is peppered with oases of settlement and greenery amidst semi-desert ochre earthy ridges, mountains and slopes, beneath the highest glacier-topped summits. Man-made Buddhist temples (gompas), prayer flags, stupas, statues and mani walls weave into the natural landscape, evidencing a unique culture that thrives and survives within an extreme environment, traditionally cut off during long winters from the outside world.

Geologically, Ladakh was once an island arc in the middle of the ancient Tethys Ocean and became caught up in the greatest continental collision of recent geological times. As well as changing Earth's geography and climate, tectonic processes have presented us in Ladakh and adjacent Kohistan with a window into deep crustal processes, allowing investigation into how the crust responds to the enormous tectonic stresses of mountain formation (orogenesis) and continental plate collision. Geological studies in Ladakh and the NW Himalayas are probably as old as settlement itself in this area. Modern scientific studies began in the late 19th and early 20th century, but became more prolific from the 1950s, and particularly from the 1970s–1980s. International and local geologists have trekked across mountains and valleys making geological maps and geoscientific observations, taking samples and determining numerous measurements from a myriad of scientific instruments, in an effort to understand the mysterious tectonic world of mountain building. These studies have revealed the impacts of magmatic and metamorphic processes from the Jurassic to the present day, rates of uplift of the mountains (up to > 1 cm/year uplift locally), original depth estimates of crust and mantle now at the surface, which were once 20-40+ km beneath our feet, the causes and character of earthquakes and their impacts, recognition of geological 'terranes' that now form a jigsaw pattern in modern geological maps, and inform us how geological units that were once far apart are now proximally juxtaposed, one to another, and evaluating how the mountains were sculpted by rivers, wind and ice.

This book offers a further development in the knowledge journey for the geology and geotectonics of Ladakh. It reports a systematic structural geological study of a number of valleys and ridges centred on the territory's capital town of Leh, and extending into remote valleys to the north-west, south-east and south. Geological and structural mapping, structural data, geological observations, recordings and analyses, alongside detailed interpretation of > 2000 photographic images, and > 100 photo-mosaic panoramas, presents the most detailed analysis to date, of the deeper basement plutonic rocks that constitute 80% of the local area's geology (the Ladakh batholith). The book takes the reader step by step through a geotectonic review and setting of Ladakh, analysis of the landscape and rock exposures and presentations of interpretations, models, hypotheses and explanations. The book is beautifully illustrated with numerous interpreted photographs and photo-mosaic landscapes, alongside colour scientific diagrams, that clearly present complex ideas and concepts, in an effort to widen access to what can be a difficult but deeply fascinating subject. I hope the reader enjoys reading as much as I have enjoyed writing the book.

Auckland, New Zealand
March 2023

Michael George Petterson

Contents

About the Author

Michael George Petterson is currently Professor of Geology, School of Science, and Pro Vice Chancellor/Dean of the Faculty of Health and Environmental Science, Auckland University of Technology, New Zealand. After completing his Ph.D. at the Department of Geology, University of Leicester, UK, in the geology and structure of the Kohistan batholith, North Pakistan, he continued his research at the British Geological Survey, University of Leicester, and the Secretariat of the South Pacific (based in Fiji). This book results from research at various intervals over the past 40 years in the NW Himalayas, particularly in north Pakistan and now NW India (Ladakh). Professor Petterson's research interests span geological mapping, structural geology, igneous geochemistry and volcanology particularly applied to island arcs, both modern and ancient (in the Pacific Islands region, UK and Caribbean) and the Ontong Java Ocean Plateau. Alongside pure geoscience, Prof. Petterson has gained much experience in senior and executive leadership within geoscience and geoscience development organisations, as well as universities, publishing widely in the application of geoscience and environmental science to development.

Professor Petterson has published over 200 papers, geological maps, book chapters and popular articles. He has worked as a guest editor for the *Geoscience Journal*, held senior leadership positions with the British Geological Survey, Auckland University of Technology and the Secretariat of the Pacific Community, and residential or long-term professional positions in Solomon Islands, Guyana, Afghanistan and Fiji.

List of Figures

List of Tables

Chapter 1
Geo-tectonic and, Topographic, and Geographical Setting of the Leh–Ladakh Region: Study Scope and Methodology

Abstract This study focuses on deformation within the Ladakh batholith and Indus Suture Rocks centred on Leh region, Ladakh Territory, NW India. The area is part of the Jurassic-Tertiary Kohistan–Ladakh-Island-Arc Terrane (KLA) which is now 'trapped' between the Eurasian and Indian megaplates within the Himalayan orogenic belt. The KLA formed as an island arc, was transformed into an Andean margin and is now within one of the world's major mountain collisional belts. Ladakh is a mountain desert/semi-desert with a high-degree of geological exposure, and a continental interior climate, experiencing cold winters and warm summers. The River Indus forms the local topographic baseline situated at heights of 3100–3300 m above sea level (asl). Surrounding mountains and ridges range up to 5500 m asl. Glacial and glacio-fluvial valleys are poorly developed south of the River Indus but form valleys up to 15 km long and 6 km wide to the north of the study region. Geological fieldwork, mapping, and structural geological studies, together with photographic image, photo-mosaic, and satellite image anlaysis, are applied to the fundamental research question: 'how has the Ladakh batholith/basement deformed during the Himalayan orogeny in the Leh–Ladakh region?'

Keywords Kohistan–Ladakh Terrane · NW Himalaya · Ladakh batholith

1.1 Broad Geo-tectonic and Geographical Contexts of This Study

This study focuses on an area of the Ladakh batholith situated in the Leh region of Ladakh NW India (Fig. 1.1). The Ladakh batholith is part of the larger Kohistan–Ladakh batholith, which extends from Afghanistan, through Pakistan, (Gilgit Baltistan) into India (Kashmir and Ladakh Union territories), with a NW–SE strike length of some 700 km, and a NE–SW width that varies between 200 and 20 km (Petterson 2010).

The Kohistan–Ladakh batholith is a component of the 2700 km Trans-Himalayan batholith extending eastwards through Tibet and into Myanmar. Figure 1.1 presents

M. G. Petterson, *Himalayan Thick-Skin Basement Deformation of the Ladakh Batholith, Leh-Ladakh Region, NW India*, Springer Geology,
https://doi.org/10.1007/978-3-031-31566-4_1

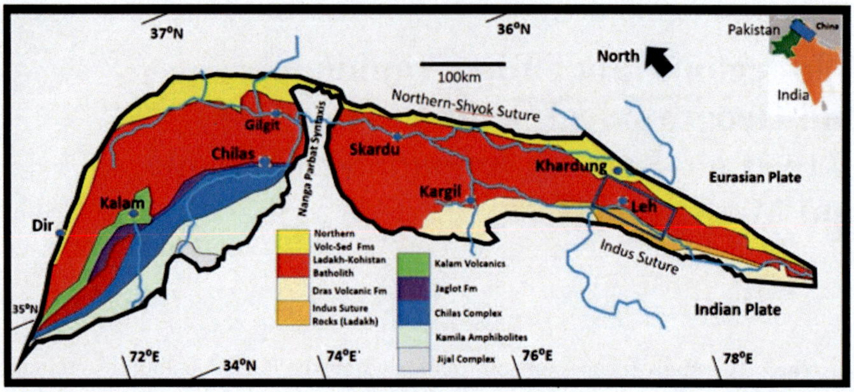

Fig. 1.1 Simplified Geological Map of the Kohistan–Ladakh Island Arc Terrane. Adapted from Petterson (2019) acknowledgement Geological Society of London. The intra-Tethys Ocean Kohistan–Ladakh Island Arc became trapped during the Himalayan Orogeny, caught between the Indian Plate to the south and the Eurasian Plate to the North. The Kohistan–Ladakh batholith occupies the largest part of the terrane and is the subject of this study, along with the Ladakh-Indus Suture rocks. The bulk of the arc terrane comprises Jurassic-Eocene aged volcanic, sedimentary and plutonic rocks, with younger Tertiary rocks forming in post-Himalayan collisional times (c. post 55–50 Ma). The blue rectangle around Leh indicates the study area

a simplified geological map of the Kohistan–Ladakh Arc Terrane (KLA). The Kohistan–Ladakh arc terrane (KLA) is bounded by two sutures: the Northern-Shyok Suture to the north and the Indus Suture to the south (Petterson 2010, 2019, and references therein) which separate Kohistan–Ladakh from the Eurasian and Indian plates respectively.

The Kohistan–Ladakh batholith forms the largest outcrop area within the island arc terrane: the batholith intrudes sequences of Jurassic–Cretaceous volcanic, sedimentary, and plutonic rocks. From c. 110–150 Ma to 55–50 Ma Kohistan–Ladakh formed an island arc and an Andean margin respectively within the Tethys Ocean, with subduction processes driving a continuous period of plutonism and volcanism, and surface processes eroding and redistributing sediment within the terrane (Petterson 2010, 2019, and references therein).

Researchers since the 1970s have differed with respect to models of the tectonic history of Kohistan–Ladakh (Fig. 1.2). It is beyond the scope of this study to explore the many details of tectonic models and readers are directed to Petterson (2010, 2019), Martin et al. (2020), Walsh et al. (2019), and Andjić et al. (2022) for further details.

All models agree that: (1) Kohistan–Ladakh was a separate island arc terrane (KLA) situated between the Eurasian Plate to the North and the Indian Plate to the south; (2) the Indian Plate separated from Gondwana in the Jurassic and drifted northwards, until its collision with Eurasia in Tertiary times; and (3) the sutures that separate the KLA from Eurasia and India (the Northern-Shyok and Indus Sutures respectively) mark the collision points and boundaries between the KLA and the Eurasian/Indian plates respectively (Fig. 1.1). Researchers debate tectonic points

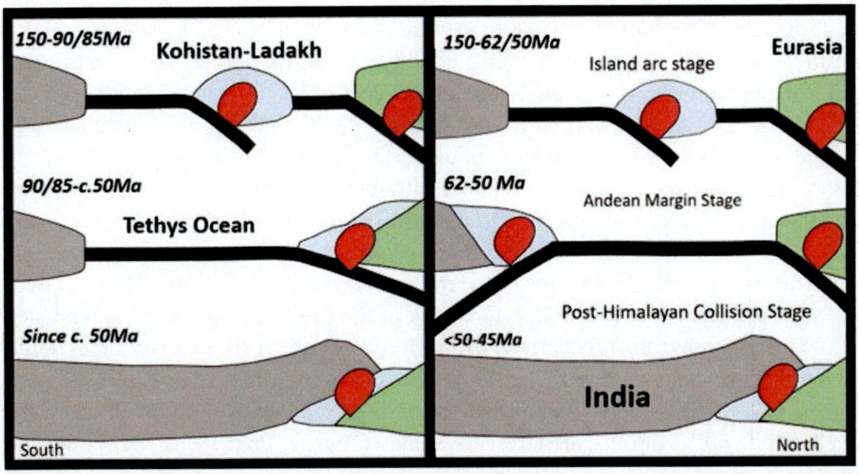

Fig. 1.2 Summary of main aspects of the tectonic history of the Kohistan–Ladakh island arc terrane (KLA). Readers are directed to the text and quoted references for detailed variations in models. Most research literature supports a model on the left-hand side of the figure. The KLA was an island arc within the Tethys Ocean from Jurassic–Cretaceous times. The KLA is then transformed into an Andean margin during the mid-Cretaceous and a continent–continent collision zone in Tertiary times. In this model the KLA collides first with Eurasia (at c. 90–85 Ma) and secondly with India (at c. 50 Ma). The right-hand side presents a summary of ideas presented by Bouihol et al (2015), Martin et al. (2020) and others, that show the KLA first colliding with India at c. 62–50 Ma and Eurasia at c. 45 Ma. Other researcher models show Tethys subduction polarity to the south, rather than the north, or subducting simultaneously to the north and south, and the presence of back-arc basins and other subduction zones. What is important for this study is that the Kohistan–Ladakh batholith (represented by the red bubbles in Fig. 1.2) has formed from magmatism in Jurassic–Miocene times within island arc, Andean-Margin, and post-continent–continent collisional times

such as: (1) the polarity of subduction of the Palaeo-Tethys Ocean, e.g. to the north or south, or both simultaneously, at different times; (2) the timings and age of the two collisions between the KLA and Eurasia/India respectively; (3) the palaeo-latitudes of the various tectonic units at different geological time periods; (4) whether the KLA collided first with Eurasia or India; and, (5) whether or not a number of arcs were simultaneously in existence at the same time within the general geographical region. Most of the research literature propose models where the earlier tectonic collision involved the Kohistan–Ladakh arc colliding with Eurasia at c. 90–85 Ma, being subsequently transformed into an Andean style continental subduction margin, with the final Himalayan continental collision beginning at c. 55–50 Ma. Kohistan–Ladakh has been, from then-on, trapped between the two mega-plates of India and Eurasia, within the Himalayan orogenic collision zone. Thus, over time, Kohistan–Ladakh experienced three different tectonic regimes: (1) as an intra-oceanic island arc regime from Jurassic: c. 150–100 Ma to 90–85 Ma; (2) as an Andean-style continental margin subduction regime (90–85 to 55–50 Ma), and (3) as a continent–continent collision regime (from 50–55 Ma to the present day). Other researchers (e.g. Martin et al. 2020) propose tectonic models that include a series of subduction zones between the Indian and Eurasian Plates and multiple-stage collisions. The Kohistan–Ladakh arc in their model collides first with India at c. 60 Ma with the 'full' continental collision beginning at 45–40 Ma. In this model the Kohistan–Ladakh arc experiences continuous intra-arc subduction, from Jurassic to c. 40–45 Ma, albeit with a tectonic change at c. 50–60 Ma as the arc collides with India. This study focuses on the deformation produced within the Ladakh batholith around Leh (Fig. 1.1) during the Himalayan collision, and aspects of the deformation within the Indus Suture rocks, which crop-out to the south of the Ladakh batholith.

Figure 1.1 presents the location of the study area centred on the capital town of Leh, Ladakh, located within NW India. For most of the time since Indian independence in 1947 Ladakh has been governed within India as part of the state of Jammu and Kashmir. This status changed in 2019, when Ladakh became a new Ladakh Union Territory. Ladakh borders China/Tibet to the north and east, Pakistan to the west, and Jammu and Kashmir Union Territory, and Himachal Pradesh, to the south. The territory is a dry mountain desert with the highest mountains reaching > 7300 m but mostly in the range 5500–7000 m, with mountain passes attaining heights of 4–5000 m (Fig. 1.3). Valley floors, dominated by the Indus and Zanskar and Shyok River systems vary in height between 2700 and 3500 m. The capital of Ladakh, Leh, is at a height of 3360 m. The region has a cold desert climate with an annual rainfall/precipitation of c. 10–12 cm per year. In Leh the average maximum July temperature is 25 °C and average minimum January temperature is – 15 °C, although much colder temperatures can be experiences at higher altitudes: the town of Drass (west of Kargil in Fig. 1.1) is reputed to be the second coldest occupied place in the world, after Siberia.

Figure 1.3 presents a summary geographical map of the study area. The River Indus flows through the centre of the area trending ENE–WSW between Phyang and Saboo and subsequently heads in a more NW–SE direction for the rest of the area. Valleys and ridges trend NE–SW, with a starkly different development style of valleys within the Ladakh batholith in the north and northeast, as compared to relatively poor valley development in the west and southwest, where the Indus Suture Rocks outcrop. Valleys within the Ladakh batholith north and east of the river Indus extend between 8 and 15 km northeast and are 3–8 km wide at their widest points. Valleys within the Indus Confluence Rocks to the west and south of the River Indus are much more constricted with valleys around Stok, Matho, and Stakna extending only c. 3–4 km into the mountains and being only 2–3 km wide. Burbank and Fort (1985) suggest that the stark differences in geological structure and lithology between the Ladakh Batholith and Indus Suture Rocks were a strong control on glacial valley development, with glaciers unable to erode deeply or make much of a northwards-progression within the Indus Confluence Rocks. Sant et al. (2011) describe the northern Indus tributary valleys to the Indus in Ladakh as 'amphitheatre valleys' and suggest that there is a relation between the size of the catchment areas around amphitheatre valleys and the extent of previous glacial and fluvio-glacial activity. The Indus floodplain is some 4–10 km wide, and comprises thick deposits of sand, silt, and gravels, but also coarser boulder deposits locally. Alluvial fan deposits, such as are present at Stok, extend the alluvial deposits into numerous valley floors, some distance from

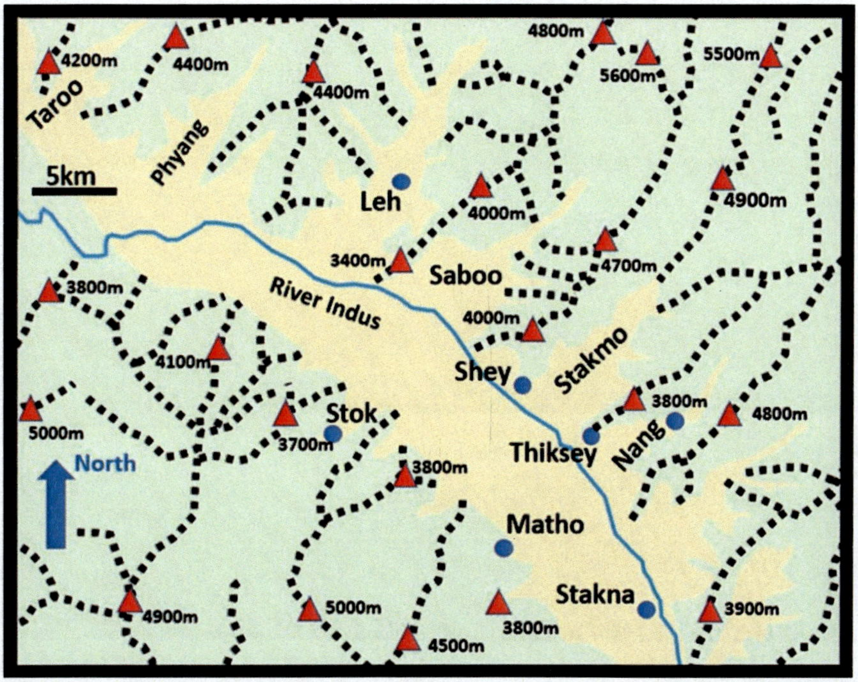

Fig. 1.3 Main Geographic Subdivisions of the Study Area. The River Indus flows WNW–ESE between Phyang and Saboo and then NW–SE between Shey and Stakna. Floodplain alluvial sands and gravels occupy the central part of the area varying in altitude from 3300 m at Stakna, 3200 m near Leh, and 3150 m SW of Phyang. The landscape has been dissected by glaciers producing valleys and ridges that predominantly trend NE–SW. Valley floors rise to over 4000 m in places like Upper Phyang. Valleys are hardly developed south of the Indus. Highest ridges reach over 5500 m in the northern part of the area. Watersheds are depicted as pecked black lines

the River Indus, as do contributions from till deposits of five different glaciation events in the area (Owen et al. 2006). The River Indus flows westwards with river heights of c. 3300 m in the SE of the study area, and 3150 m in the west of the study area. Ridges rise quickly from the valley floor attaining heights of over 5500 m in the northern parts of the study area, although mostly between 4000 and 5000 m high. Permanent glaciers and snow fields are locally present above 4500 m: these are small glaciers at the current time. Large U shaped valleys with extensive, poorly sorted, thick, till deposits, terminal moraines, side moraines, eskers, and varve clay deposits are evidence of recently active glacial and periglacial processes. Figures 1.4, 1.5, 1.6, and 1.7 present examples of the topography and geography of the study region.

Topography is strongly controlled by geology, structure, active and recent tectonics, and surface erosional and depositional processes related to rivers, glaciers, and wind. The Ladakh batholith tends to form more rounded mountains with a greater development of valleys (as discussed above). Indus Suture Rocks produce

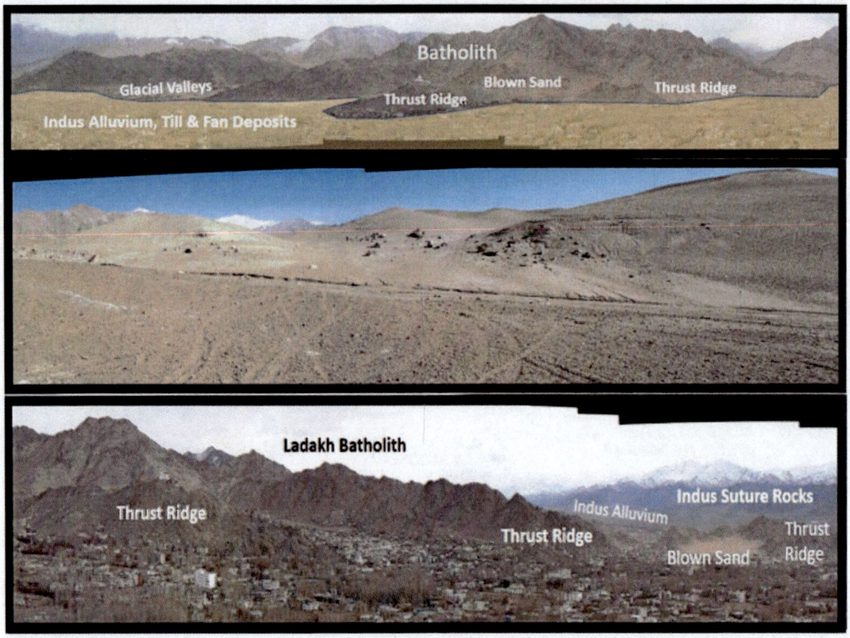

Fig. 1.4 Examples of variations in topography in the Leh–Ladakh Region. Uppermost Photomontage (Leh area) shows the nature of the Ladakh batholith mountainous topography with ridges and intervening amphitheatre glaciated valleys. Two outlier ridges, one with the Shanti Stupa on top are examples of thrust ridges formed from geological thrusting and now isolated from their parental ridges. The lower plains are dominated by Indus alluvium, alluvial fans, and till deposits. The central photomontage (Phey area, lower Phyang) is an example of a typical arid Ladakh desert landscape, with extensive sand deposits, locally forming dunes and rounded hills. The lower (Leh area) photomontage displays Indus Suture Rocks on the right, Ladakh batholith rocks on the left, and the intervening Indus Alluvium. Blown sand occupies the current golf course within a natural intra-montane bowl feature. Thrust Ridges are present including one (centre left) on which the Tsemo Gompa and Leh Palace are built

sharper, more rugged mountain topography, and a relative paucity of valleys. Alluvial fans are common, and form steeply inclined conical mounds composed of poorly sorted, angular to sub-angular blocks, boulders, gravel, sand, and silt. Alluvial fans around Stok and Matho extend up to 10 km from a steep, mountain-hugging source region outwards towards the River Indus, eventually merging with the Indus alluvial deposits (Fig. 1.6). Indus alluvial deposits are thick and extensive, tending to form a blanket of silt, sand, blocks, and boulders which buries topography (Figs. 1.4 and 1.6). The Indus is a highly active river, rejuvenated by the uplift of the Himalayas, and currently erodes through previously deposited alluvium and palaeo-floodplains, forming incised meanders, incised channels, and deep gorges (Fig. 1.5). Owen et al. (2006) proposed a model, based on cosmogenic dating and glacial deposit studies, for the existence of five different glaciation events within the Leh–Ladakh region,

Fig. 1.5 The tectonically rejuvenated Zanskar River cuts deeply through the Indus Suture and older rocks of the Indian Plate south of the River Indus. Valleys south of the Indus are limited mainly to areas cut by larger rivers and are gorge-like in form

dating from 430,000 years to 6000 years, with each glaciation event being progressively less extensive. In valleys such as Phyang, Leh, Saboo, and Stakmo the effects of cumulative deglaciation events are clearly visible in terms of valley incision, and the presence of moraine, till, and glacio-fluvial deposits (Fig. 1.6). Individual moraines/glacio-fluvial deposits can extend up to 4 km in length and form mounds of sand and gravel up to 100 m above the valley floor. Blown sand deposits are common and accumulate within intra-montane basins (e.g. Fig. 1.4), plastered onto craggy mountain ridge slopes (Fig. 1.6) or forming thicker, more extensive dune beds (Fig. 1.4).

1.2 Research Objectives and Methodology of This Study

The central focus of this study is to answer the research question: 'how did the Ladakh batholith respond to deformation caused by the Himalayan Orogeny?'

To answer this question a number of related research questions need to be asked: (1) is it possible to spatially and systematically map deformation within the Ladakh batholith?; (2) is it possible to geologically and structurally map the Ladakh batholith,

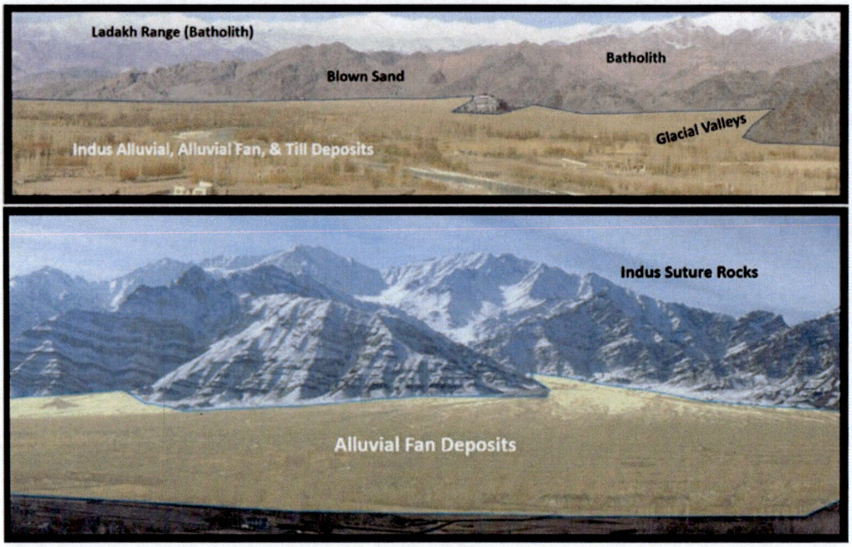

Fig. 1.6 Upper photomontage (from Stakna looking NW) illustrates the broad relatively flat morphology of the Indus Alluvium. The batholith forms lower and higher mountains of the Ladakh Himalayan Range, with snow-capped mountains attaining heights of over 5500 m. Blown sand is plastered on some mountain slopes. Lower photomontage (Matho from Stakna) presents two coalescing alluvial fans which form steeper slopes in the upper valleys, merging with the Indus Alluvium within the Indus valley

such that value is added to the crustal growth and development model of the batholith, and this in turn can reveal further insights into batholith deformation?; (3) Can the spatial mapping of deformation be used to develop a spatial model of deformation across the batholith?; (4) Is it possible to document the rheological and 3-D rock fabric changes required to produce extensive compressional brittle deformation?; (5) Can the brittle–ductile transition zone be identified spatially within the Leh–Ladakh region, and how does deformation manifest itself in the brittle, brittle–ductile transition, and ductile zones? Additionally, the gross structure of the Indus Suture Rocks in the area were analysed, which provides another lens to view deformation in the Leh–Ladakh area.

The methodology applied in this study is field-based structural geology to which is added image analysis primarily from an extensive photograph collection made in the field and analysis of satellite imagery. Fieldwork was undertaken in two field seasons: the Ladakh winters of 2018 and 2019 between November and January. Systematic field traverses were made from the Taroo–Phyang–Phey region in the NW through the Leh, Saboo, and Stakmo valleys, and parts of the Indus valley and valleys in the south and east of the region. Additional trips were made across a wider area, including the Zanskar River, Gangles, Chemrey, Likir and Sakti areas. Field equipment consisted of a compass-clinometer, hand lens, metric ruler, GPS, and Fuji-Film XP camera. Systematic field observations and measurements were recorded

Fig. 1.7 The northern valleys of the Leh–Ladakh region contain abundant evidence of glacial and glacio-fluvial processes. The Upper photomontage (lower-mid Phyang) presents the Phyang Gompa, and numerous stupas built on top of a terminal glacial moraine. The central photomontage is from Upper Phyang looking south towards the Indus Suture Rocks, presents a southward trending poorly-sorted linear moraine deposit within a wide valley floor. The lower photomontage (upper Leh valley) presents a linear moraine with an esker-like morphology that extends some 4 km in length and produces a feature up to 100 m high locally

outcrop by outcrop, and general open panorama observations, including location, lithology, geology, topography, bedding, cleavage, schistose foliation, gneissic foliation, fracture styles and geometries, and shear deformation styles and geometries. A photographic archive of over 2000 images was taken from outcrop to open-view panorama scales. The images have been systematically archived and described. Many photographs were taken with the intention of producing photo-mosaics that allowed for synthetic structural analysis: Microsoft's *Image Composite Editor* software was used to create photomontages (Microsoft 2022). Over 100 separate photomontages were created and analysed as a key method for this study. Structural data were plotted as stereograms and rose diagrams using the Georose plotting software package (Yong Technology 2022).

Sections below address all aspects of the research questions above. A final section presents broader perspectives and models relating to the deformation of the Ladakh Batholith. No study operates within a vacuum. The following section presents a broad overview of the geology, structure and tectonics of the Ladakh region and adjacent areas to provide background and context for this study.

References

Andjić G, Zhou R, Jonell TN, Aitchison JC (2022) A single Dras-Kohistan-Ladakh arc revealed by volcaniclastic records. Geochem Geophys Geosyst 23:e2021GC010042

Bouihol P, Schmidt MW, Burg J-P (2015) Magma transfer and evolution in channels within the arc crust: the pyroxenenitic feeder pipes of Sapat (Kohistan, Pakistan). J Petrol 56(7):1309–1342

Burbank D, Fort M (1985) Bedrock control on glacial limits: examples of the Ladakh and Zanskar ranges, North-Western Himalaya, India. J Glaciol 31(108):143–149

Martin CR, Jagoutz O, Upadhyay R, Royden LH, Eddy MP, Bailey E, Nichols CIO, Weiss BP (2020) Paleocene latitude of the Kohistan–Ladakh arc indicates multistage India-Eurasia collision. Proc Natl Acad Sci 117(47):29487–29494. https://doi.org/10.1073/pnas.2009039117

Microsoft (2022) https://microsoft-image-composite-editor.en.softonic.com/. Image Composite Editor

Owen LA, Caffee MW, Bovard KR, Finkel RC, Sharma MC (2006) Terrestrial cosmogenic nuclide surface exposure dating of the oldest glacial successions in the Himalayan orogen: Ladakh Range, northern India. Geol Soc Am Bull 118(3–4):383–392. https://doi.org/10.1130/B25750.1

Petterson MG (2010) Review of the geology and tectonics of the Kohistan island arc, north Pakistan. In: Kusk TM, Zhai M-G, Xiao W (eds) The evolving continents: understanding processes of continental growth, vol 338. Geological Society, London, Special Publications, pp 287–327

Petterson MG (2019) The plutonic crust of Kohistan and volcanic crust of Kohistan-Ladakh, North Pakistan/India: lessons learned for deep and shallow arc processes. In: Treloar PJ, Searle MP (eds) Himalayan tectonics: a modern synthesis, vol 483. Geological Society, London, Special Publications. https://doi.org/10.1144/SP483.4

Sant DA, Wadhawan SK, Ganjoo RK, Basavajah N, Sukumaran P, Bhattacharya S (2011) Morphostratigraphy and palaeoclimate appraisal of the Leh Valley, Ladakh Himalayas, India. J Geol Soc India. 77:499–510. https://doi.org/10.1007/s12594-011-0057-9

Walsh JMJ, Buckman S, Nutman AP, Zhou R (2019) Age and provenance of the Nindam Formation, Ladakh, NW Himalaya: evolution of the intraoceanic Dras Arc before collision with India. Tectonics 38:3070–3096. https://doi.org/10.1029/2019TC005494

Yong Technology (2022) http://www.yongtechnology.com/georose/. Georose package. Accessed throughout 2022

Chapter 2
Overview of the Geology, Structure, and Tectonics of the Ladakh Batholith and Indus Suture Rocks

Abstract The Kohistan–Ladakh batholith forms the most voluminous geological unit within the Kohistan–Ladakh island arc terrane, constituting c. 80% of the outcrop area in the Ladakh region. The batholith is up to 30 km thick, with emplacement depth estimates of various components of the batholith ranging between 10 and < 2 kb (c. 35–7 km depth): in Ladakh estimates are > 4 to c. 2 kb (c. 15–7 km depth). The Ladakh batholith is heterogeneous and complex, comprising plutons, stocks, dykes and sills that range in composition from hornblende cumulates to felsic rich leucogranites. The proportion of lithological compositions exposed in the batholith is c. 10–20% mafic/ultramafic, 20–30% intermediate, 50% acid, and 10% leucocratic-felsic acid sheets. Primary mineralogy is dominated by hbl-biot-plag-qz-alk feld-with minor px. Minor mineralogy comprises zirc-sph-ilm-mag-py-cp-xen. Batholith age estimates vary between 150 and < 30 Ma, although mostly 120 Ma or less, and 78–45 Ma in the Leh-Ladakh region. Metamorphic grade varies from very low to amphibolite grade, with sericite-lower greenschist grade being most common. Non-batholithic rocks in Ladakh comprise the Late Triassic–Cretaceous clastic-limestone marine Lamayuru Complex, the Jurassic–Cretaceous Dras Volcanics, the Palaeocene–Eocene marine-sedimentary Tar Formation, including the youngest marine Nummulitic limestone, and the Eocene-Upper Miocene Indus Suture molasse sequence. Deformation has produced central upright, southerly south-verging, and northerly north-verging thrusts and folds, with deformation from c. 70 Ma to the Pliocene, and modern active tectonics. Low temperature geothermometry suggests Ladakh uplifted through the 150 °C isotherm between 30 and 8 Ma (youngest ages in north Ladakh).

Keywords Ladakh batholith · Indus Suture molasse · Geochronology and stratigraphic age

M. G. Petterson, *Himalayan Thick-Skin Basement Deformation of the Ladakh Batholith, Leh-Ladakh Region, NW India*, Springer Geology,
https://doi.org/10.1007/978-3-031-31566-4_2

2.1 Overview of the Kohistan–Ladakh Batholith

Chapter 1 and Fig. 1.1 presented a broad outline of the Kohistan–Ladakh Island Arc Terrane (KLA). There are numerous research papers concerning the KLA including comprehensive reviews of the terrane by Petterson in 2010 and 2019. It is beyond the scope of this work to provide a detailed overview of the terrane, and instead there is a focus on a synthetic analysis of the key contextual data and ideas required for this study. This analysis includes data and models relating to the age and lithological character of the Ladakh batholith, the Indus Suture Rocks, and studies of the structural geology and deformation of these rock sequences, as part of the greater terrane.

Figure 1.1 presents the spatial distribution of the key stratigraphic units of the KLA. In Kohistan a particularly complete section through an island arc is exposed: perhaps the best section through an obducted arc in the world. From the Indus Suture northwards the geological units include: (1) the ultramafic, plutonic, Jijal Complex, metamorphosed to granulite and amphibolite grade; (2) the meta-basic volcanic and plutonic rocks of the Kamila Amphibolite Belt; (3) the large sheet-like, basic, Chilas Complex; (4) the meta-sedimentary, clastic Jaglot Group and volcanic-sedimentary Kalam Volcanics; (5) the Kohistan batholith; and (6) a sequence of volcanic and sedimentary units that forms both the northern part of Kohistan and Ladakh, adjacent to the Northern-Shyok Suture. These volcanic rocks are basaltic to andesitic, locally rhyolitic, and are succeeded, and interbedded with limestone and clastic sedimentary units. In Ladakh the sequence is thinner and less complete than Kohistan, comprising from south to north: (1) the basic-intermediate Dras Volcanics and younger Indus Suture Rocks; (2) the Ladakh Batholith and (3) Northern volcanic-sedimentary units. Ophiolitic sequences and melanges are present in both Ladakh and Kohistan associated with the two sutures and marking evidence for obduction tectonics linked to the collision of the arc with the Eurasian and Himalayan Plates. Radiometric and stratigraphic age data span Jurassic-Mio-Oligocene, with the bulk of the arc system forming in Jurassic–Cretaceous times (e.g. Andjić et al. 2022).

Figure 2.1 presents a generalised summary of the lithological character of the Kohistan–Ladakh batholith based on the work of many of the researchers cited in this text, alongside new work presented in this study. The batholith intrudes pre-existing arc crust composed of predominantly basalts, and basaltic andesites, with cherts, limestones, and greywackes, but also andesites, rhyolites, and plutonic rocks. The batholith is dominated by diorite and granodiorite with associated plutonic rocks that include ultramafic amphibole-rich hornblendite rocks, gabbros, granites, basic dykes, and leucogranitic sheets, dykes, and pegmatites. The dominant mineralogy is hornblende-biotite-plagioclase-quartz-K-feldspar, with pyroxene present in addition to hornblende (and often reacting to form hornblende) in the gabbros and gabbroic-diorites. Zircon-sphene-ilmenite-magnetite-pyrite-chalcopyrite-xenotime are common minor phases. Basic to intermediate dykes cut the larger plutonic bodies throughout the evolution of the batholith, although in some areas they are more common during the earlier and later batholith stages. In many parts of the batholith the older plutonic rocks are gabbros, gabbroic diorites, and diorites, cut by younger granodiorites and granites with acid sheets/leucogranites and pegmatites forming the latest stages of magmatism. However, these relations are not ubiquitous, with co-eval basic and more evolved plutons present at outcrop levels.

Geological surveys and other research activities that have had a focus in the KLA and surrounding areas span > 100 years. The earliest pioneer geologists include Desio (1959, 1964) Gansser (1964, 1979, 1981), Hayden (1914), Ivanec et al. (1956), and Wadia (1932). The recognition of Kohistan–Ladakh as an island arc dates back to the 1970s and 80s including researchers such as, Bard et al. (1980), Bard (1983), Reuber (1986, 1989), Thakur (1981), Tahirkheli (1979), and Tahirkheli and Jan (1979). Researchers that built upon the island arc model and had a particular interest in geochemical, isotopic, magmatic, structural, and field-related studies of the Kohistan–Ladakh batholith include Ahmad et al. (1998), Andjić et al. (2022), Bignold and Treloar (2003), Bignold et al. (2006), Honegger et al. (1982), Clift et al. (2002, 2005), Coward et al. (1982a, b, 1986, 1987), George et al. (1993), Jagoutz (2009, 2010, 2014), Jagoutz and Schmidt (2012), Jagoutz and Keleman

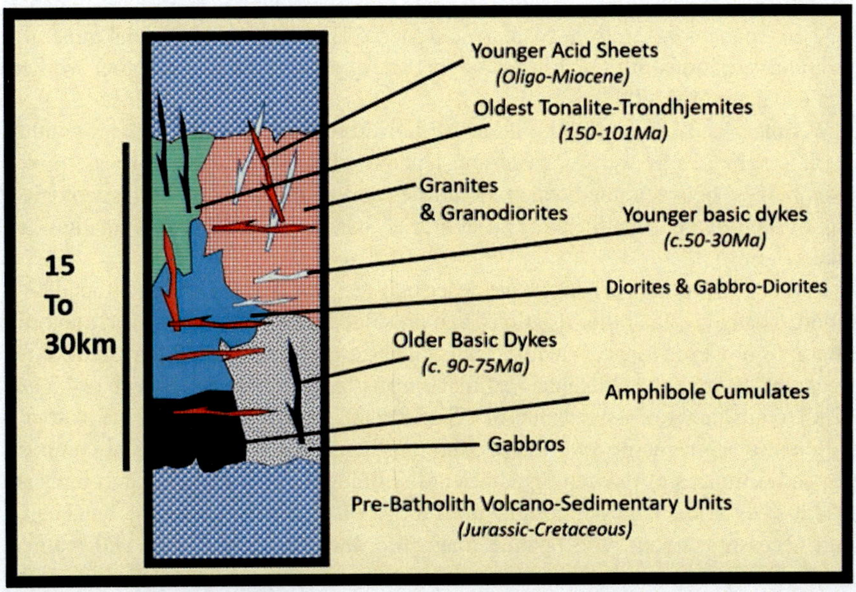

Fig. 2.1 Lithological section through the Kohistan–Ladakh Batholith. The batholith intrudes pre-existing arc volcano-sedimentary crust. The full plutonic compositional suite from gabbro, gabbro-diorite, to granodiorite and granite is present within the batholith. In many parts of the batholith basic rocks are intruded by more silica rich rocks, although this trend can reverse in time also. Cumulates rich in amphibole are common as discrete bodies and xenoliths within other plutons. Basic dykes of different ages intrude the batholith, some at earlier, and some at latest stages. Later acid sheets that cut all other rocks are present in many parts of the batholith. Whilst most batholiths are 5–17 km thick, exceptionally they may reach 30 km. See Petterson (2019, 2010), Martin et al. (2020), Jagoutz (2009), Jagoutz and Schmidt (2012) and Jagoutz and Keleman (2015) for further details

(2015), Jagoutz et al. (2011, 2019), Khan and Coward (1991), Khan et al. (1997, 2009), Kumar (2008, 2010a, b, 2020), Kumar and Singh (2008a, b), Kumar et al. (2012, 2016), Martin et al. (2020), Petterson and Windley (1985, 1986,1991, 1992) and Petterson et al. (1991a, b, 1993), Petterson and Treloar (2004) Ravikant et al. (2009), Saktura et al. (2020), Schaltegger et al. (2002), Thanh et al. (2010), Treloar et al. (1989, 1996), Upadhay et al. (2008), Weinberg and Dunlap (2000), Zeitler and Chamberlain (1991). For reviews of the Kohistan–Ladakh arc see Petterson (2010, 2019).

The field, geochemical, and petrological results published from the above researchers have a high degree of convergence summarised in Fig. 2.1. The Kohistan–Ladakh batholith is dominated by intermediate-granodiorite plutonic rocks, with grain sizes that vary between medium-coarse grained, and accompanied by more basic gabbroic and more acid granitic-leucogranitic rocks.

Honegger et al. (1982) describe the batholith around Kargil and east of Leh being comprised of 10–20% mafic rocks, 20–30% diorites, 50% granitoids, and 10% leucogranitic sheets. Mineral modes vary between: Pyroxene 2 to 3–0%; Hornblende 35–1%; Biotite 15–1%; Plagioclase: 60–30%; Quartz: 7–35%; K-feldspar: 3–32%; Opaques 9–5% from basic to acid rocks. These variations, modal mineralogies, and variations in modal mineralogies are quoted by numerous other workers (e.g. Ravikant et al. 2009).

Weinberg and Dunlap (2000) describe dioritic plutons intruded by later granites and leucogranites as well as basic and leucogranitic dykes. They discuss 'mixed assemblages' between tonalites, granodiorites and granites and present a geological map of the Phyang-Saboo area centred on Leh which also includes migmatites and dyke swarms.

Kumar and workers (op.cit.) have described sections of the batholith around Leh, Saboo, Chang La, and other locations, and presented field photographs and photomicrographs of various aspects, with a focus on basic-acid magma mingling and related processes. Kumar and colleagues estimate intrusion pressures of between c. 4.3 and 2.2 kb (equivalent to crustal depths of 12–15 km to 7–9 km) and classify the plutonics as ilmenite or magnetite series depending on their opaque mineralogical composition and estimates of oxygen fugacities at the times of fractionation. Granitoids are described as assemblages of hornblende, biotite, plagioclase, K-feldspar, and quartz, with accessory zircon, and opaque magnetite and ilmenite. Kumar and workers emphasise the continual co-existence of basic and acid magmatism throughout the magmatic history of the Ladakh batholith. Basic magmatic enclaves forming 35–65% of the crystal content of inter-mingled basic-acid magmas form dark xenoliths within an acid host, with limited chemical or magmatic interaction between the two magma types. They argue that basic dykes are forcibly broken-up by strong magma flowage and convection currents within original magma chambers. Xenoliths may be locally aligned to magma flowage currents or can be randomly aligned. Xenolith shapes vary between rounded and highly angular and irregular. Reaction rims are rare.

Fig. 2.2 Images from the Kohistan Batholith. **a** Presents late stage, c. 29–34 Ma (Rb–Sr, Petterson and Windley 1985) suite of leucogranite dykes that cut a gabbro-diorite intrusion. **b** Presents the Matum Das trondhjemite-tonalite dated at c. 102 Ma (Petterson and Windley 1985, Rb–Sr) and 150 Ma (Jagoutz et al. 2019, U–Pb zircon) intruding the dark Chalt Volcanics (stratigraphic estimated age as Cretaceous (Pudsey 1986; Pudsey et al. 1985a, b, 1986) and intruded by the Jutal Das basalt dykes (dated at 75 Ma, Ar–Ar, Treloar et al. (1989): red lines highlight two dykes from the dyke swarm. **c** Presents the Shirot granite-leucogranite (dated at 40 Ma, Rb–Sr, Petterson and Windley (1985) which intrudes Cretaceous volcanic rocks, white pecked line indicates the intrusive margin. The Matum Das intrusion is postulated as representing the earlier stage 1 batholith, the Shirot, Jutal dykes, and gabbro-diorite (**a**) postulated as stage 2 and the leucogranite sheets as stage 3 (Petterson and Windley 1985, 1986, 1992). Jagoutz and Schmidt (2012) prefer a continuous magmatism model for Kohistan

Petterson, Jagoutz, George, Bignold, Khan, Treloar and co-workers have made extensive field observations in the Kohistan batholith. There are many similarities with descriptions above. Some possible differences between the Kohistan and Ladakh regions of the batholith include: (1) a higher abundance of hornblende ultramafic rocks in the form of xenoliths, large to very large pods within plutons, and high-volume hornblendites; (2) local high densities of basalt dykes; (3) the presence of older tonalite-trondhjemite plutons (e.g. Matum Das) and (4) high density of acid sheet intrusions in parts of the batholith, which can form up to 50% of local outcrops. Petterson and Windley (1985, 1986, 1992) and Petterson (2010, 2019) summarise the Kohistan batholith as a unit of complex and varied plutonics with numerous individual plutons, stocks, sheets, and dykes, with larger intrusions ranging in size from 1 km to 5–10 km in diameter. Some intrusions have clear cross-cutting margins with volcanic or sedimentary country rock. Compositions vary from homogeneous to heterogeneous and basic to acid (granite, trondhjemite, granodiorite, diorite, and gabbro). Abundant leucogranitic sheets are present locally. Crystal sizes vary from medium-coarse and rock textures from granular-sub-granular to porphyritic. Mineralogies include hornblende-pyroxene-biotite-plagioclase-alkali feldspar-quartz-magnetite-ilmenite-zircon-sphene-xenotime. Hornblende xenoliths, pods, layers, and ultramafic bodies are common. Xenoliths (sedimentary, hornblende-rich, lithic) are common, many of which are stretched. Leucogranitic sheets can locally form 30–40% of outcrops. Jagoutz and Schmidt (2012) report individual intrusions dominated by quartz-monzonite to granodiorite. Granitoid-diorite dykes up to 100 m wide form complex cross-cutting relationships and display many erosional and magmatic flow features in the southern part of the batholith, similar to those reported by Kumar and colleagues for Ladakh (op.cit.). Petterson and Windley (1985, 1986, 1992) proposed a three-stage magmatic model for the Kohistan batholith: (1) pre-85–90 Ma linked to the island arc tectonic stage; (2) 85–90 Ma to 40 Ma linked to the Andean stage and (3) post 45 Ma linked to the post-collisional stage. Jagoutz and Schmidt (2012) prefer a continuous-magmatic model for the batholith, with proposing that three geochemical groups exist within the batholith including granitoids broadly similar to the Archaean Tonalite—Trondhjemite—Granodiorite assemblage. Kohistan magmas can be modelled as hydrous fractional crystallisation from a basaltic parent at different pressures within a subduction-arc environment. Cumulus fractionating phases are present as cumulates within the batholith.

The batholith is variably metamorphosed. Many workers report the presence of sericite and clay minerals, epidote, chlorite, and actinolite. Jagoutz (2014) and determined paleo-pressures across Kohistan of between > 10 and 2 kb. The highest pressures were determined from rocks close to the rapidly uplifted Nanga Parbat Syntaxis and the lowest pressures estimated from rocks in mid-northernmost Kohistan (Fig. 1.1). The Kohistan batholith intruded between depths of 8–9 km and 25–35 km with a complete batholith-crustal section exposed within Kohistan. This study reports the presence of amphibolite grade rocks locally in the Leh region. Metamorphic grades within the batholith vary mainly from low sericite grade to upper greenschist grade with higher-grade amphibolite to granulite and partial-melt grade locally present.

Table 2.1 presents a summary of published age dates of the Ladakh batholith, east of the Nanga Parbat Syntaxis. The oldest magmatic rock yet dated in Ladakh is the 159–152 Ma Changmar Complex in the Shyok Suture Zone region. Satkura et al. (2020) suggest that this may be similar to the oldest dated unit in Kohistan (the Matum Das Tonalite-Trondhjemite) although compositionally they are quite different. However, structurally and chronologically they may represent some of the older parts of the plutonic sequences in the Kohistan–Ladakh (K–L) batholith. Jadoon et al. (2019) dated the oldest definitive Ladakh Batholith rock at 121 Ma followed by Honegger et al. (1982) at 102 Ma. Kumar et al. (2016) have dated the Tirit Granite at 105–109 Ma but this may be part of a different batholith to the K-L Batholith. A range of ages are reported of between 78 and 60 Ma from Somau (SW of Kargil) Hunder (Nubra area), and Leh-Chang La, with younger < 60 to 45 Ma ages from Shey, west Leh area, Leh-Chang La-Darbuk areas, and Khardung La (equivalent to the co-equal Khardung Volcanics). There may be two episodes of magmatism within the batholith: (1) pre-100 Ma; and (2) 78–45 Ma. The Kohistan batholith has age ranges of 150-c. 29 Ma with very young ages of c. < 1 to 5 Ma associated with the Nanga-Parbat Syntaxis exhumation (e.g. George et al. 1993).

Age data from the Ladakh batholith indicate that dioritic and granitic magmatism were co-eval within the 78–45 Ma magmatic stage (Fig. 2.3). This study suggests that published data have omitted results from detailed field relation analysis which can target a wider range of sampling to elucidate a broader range of ages and magmatic sub-stages. In many parts of the Leh area the oldest plutonic 'basement', which itself intrudes earlier arc crust, is invariably gabbro-diorite ± associated high grade metamorphic basement rocks. Similarly, the many acid sheets, e.g. in Stakmo valley, themselves present a complex history which can be further elucidated by targeted sampling and age dating.

2.2 Overview of the Indus Suture Rocks

The Indus Suture Rocks of Ladakh constitute a series of geological units that crop out between the Indus Suture and the Ladakh batholith (Fig. 1.1). Numerous researchers have described and defined these sequences. There are a rather confusing series of stratigraphic terms in the literature. It is beyond the scope of this study to untangle the complexity of the various stratigraphic definitions: instead this study presents a simplified version of the stratigraphy focusing almost entirely on the youngest Indus Group Molasse sequence which are of most relevance here. Readers are directed to Brookfield and Andrews-Speed (1984a, b), Clift et al. (2002), Garzanti and Van Daver (1988), Henderson et al. (2010, 2011), Jain et al. (2009), Searle (1983, 1986) Searle et al. (1988, 1990, 1992, 1997), Sinclair and Jaffey (2001), Van Haver (1984) Wu et al. (2007), Thakur (1981), Walsh et al. (2019), and Zhou et al. (2020) for further details.

Searle et al. (1988) and Thakur (1981) describe three key units between the Ladakh batholith and the Indus Suture: The Lamayuru Complex, the Dras Volcanics and associated sedimentary rocks, and the Indus Molasse Group. The Lamayuru Complex is Late-Triassic-Jurassic to Cretaceous in age, comprising a thickness of c. 1000–1500 m of shales, siltstones, graded sandstones, and limestones, with many units displaying turbiditic characteristics. The sequence has been extensively thrusted towards the SW with later N-directed backthrusts affecting the sequence locally. The Dras Volcanics comprises a series of outcrops spread across Ladakh, of Jurassic–Cretaceous age, dominated by basalt-basaltic andesite lavas and tuffs, with minor andesites, and rhyolites. These are interbedded with, and intercalated within, various thicknesses of clastic sedimentary rocks (see Petterson 2019 and references therein for an in-depth analysis of the Dras Volcanics and related units across the Kohistan-Ladakh terrane).

This study focuses upon the lithological character and sedimentological facies interpretations of the Indus Molasse or Indus Group (Fig. 2.4). Searle et al. (1990)

Table 2.1 Summary of radiometric isotopic ages from the literature for the Ladakh Batholith, east of Nanga Parbat

Age	Method	Lithology	Location	References
58.4 ± 1 Ma 49.8 ± 0.8 Ma	U–Pb Zircon (Zr)	Diorite-Granite intruded by leucogranite dykes	Saboo, east of Leh	Weinberg and Dunlap (2000)
70–50 Ma	U–Pb (Zr) and Ar–Ar	Diorite-Granite	Leh-Phyang-Saboo area	Weinberg and Dunlap (2000) general analysis
61.5 ± 2 Ma	U–Pb (Zr)	Mixed Plutonic Rock	SE of Leh on the way to Shey	Weinberg and Dunlap (2000)
46 ± 1 Ma	Ar–Ar	Mafic Dyke Suite	Phyang, west of Leh	Weinberg and Dunlap (2000)
121–45 Ma	U–Pb (Zr)	One gabbro (68 Ma), four granites	Deosai Plateau, S-SW of Skardu, Pakistan: Deosai Volcanics 123–48 Ma (U–Pb (Zr))	Jadoon et al. (2019)
102 Ma	U–Pb	Granodiorite	Kargil	Honegger et al. (1982)
78–44 Ma	K–Ar (biotite)	Granodiorite	Somau, SW of Kargil	
45–49 Ma	K–Ar (Biot) and Rb–Sr (WR)	Granite and Granodiorite	Shey, east of Leh	
48.7 ± 1.6 Ma	K–Ar (Biot)	Granodiorite	30 km east of Shyok	
67 ± 2 to 3 Ma	U–Pb (Zr)	Diorite-Granite	Hunder, Nubra, North Ladakh	Upadhay et al. (2008)
45–53 Ma		Granodiorite and Diorite	West of Leh, Indus	
58 Ma		Granodiorite	Chang La, north of Stakna, East of Leh	
50–68 Ma	U–Pb (Zr)	Granite 68 Ma), Granodiorite (64 Ma) and Diorite (50–51 Ma)	SE of Leh to Chang La to Darbuk	Ravikant et al. (2009)
159–152 Ma	U–Pb (Zr)	Gabbro	Changmar Complex Shyok Suture area	Saktura et al. (2020)
49–51 Ma	K–Ar (Biot)	Granite-Granodiorite	Khardung La	Thanh et al. (2010)
61–64 Ma	K–Ar (Biot)	Diorite to Granite	Hunder, Nubra	
105–109 Ma	U–Pb (Zr)	Granite	Tirit, Shyok Suture	Kumar et al. (2016)
67 ± 1.1 Ma	U–Pb (Zr)	Granite	Hunder, Nubra	

(continued)

Table 2.1 (continued)

Age	Method	Lithology	Location	References
103–45 Ma	Various	Various	Ladakh Batholith	Kumar et al. (2016); general analysis of range of ages

Earliest magmatism in Ladakh is dated at 159 Ma (Changmar Complex, Shyok Suture area). The Ladakh batholith is dated between 102 and 45 Ma. This compares with published radiometric ages from the Kohistan Batholith of c. 150 to < 30 Ma

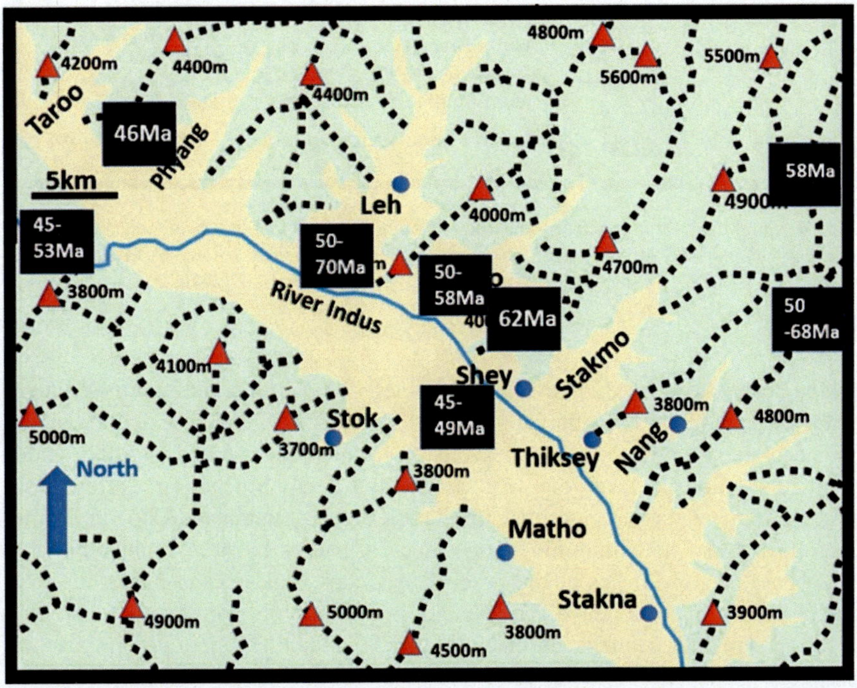

Fig. 2.3 Radiometric ages of the Ladakh Batholith in the Leh region. The data are consistent with an age range of 70–45 Ma. Most ages are between 45 and 62 Ma

describe the development of fore-arc and intra-continental intra-montane basins some 2000 km long and c. 60 km wide that existed from Eocene to Oligo-Miocene times. Several researchers have proposed a range of stratigraphic units that comprise Indus Suture Rocks formed in collisional-post-collisional times, as the Tethys Ocean finally closed and non-marine sedimentation became predominant. This account takes particular note of the Sinclair and Jaffey (2001) stratigraphy as they describe the units in detail from a range of localities. Other particularly notable references include Thakur (1981), Brookfield and Andrews-Speed (1984a, b) Searle et al. (1990), Clift et al. (2002), Wu et al. (2007), Zhou et al. (2020), and Henderson et al. (2010, 2011).

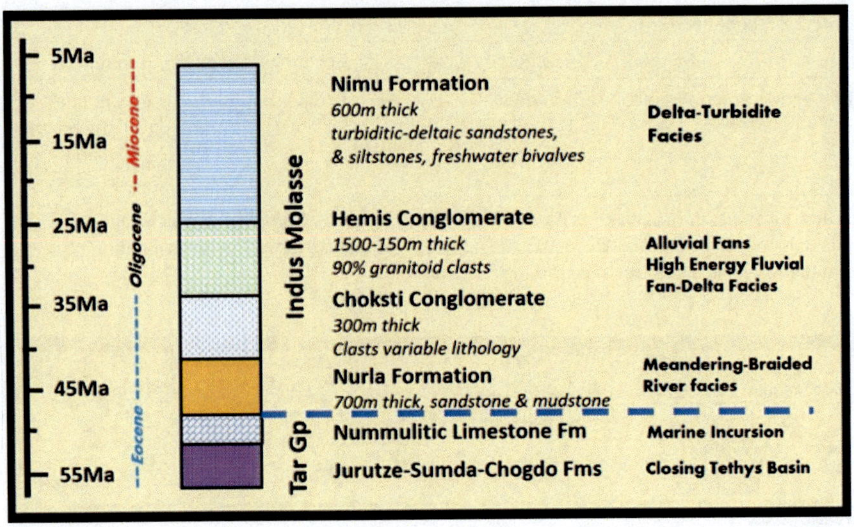

Fig. 2.4 Simplified Stratigraphy of the Indus Suture Rocks. Adapted from range of authors including Sinclair and Jaffey (2001), acknowledgement, Geological Society of London, Searle et al. (1988, 1990), Jain et al. (2009), Wu et al. (2007), Henderson et al. (2010, 2011), and Thakur (1981)

The Eocene-Miocene Indus Molasse is some 2–3000 m thick and records sedimentation within alluvial fan, alluvial delta, turbiditic, lacustrine, high-medium energy fluvial, braided-fluvial, and fluvial overbank facies (Figs. 2.5 and 2.6). The Indus Molasse overlies the Tar Group (e.g. Searle et al. 1990; Sinclair and Jaffey 2001) which comprises the Nummulitic Limestone, Jurutze, Sumda and Chogdo Formations (Fig. 2.4). The Nummulitic Limestone Formation, biostratigraphically dated at Eocene (c. 50 Ma) lies at the base of the Indus Molasse and marks the final marine incursion into Ladakh. The Nummulitic Limestone Formation comprises a c. 100–180 m thick sequence of black-dark grey crystalline limestones, mudstones, sandstones, and pebbly conglomerates. The Nummulitic Limestone Formation overlies the Jurutze, Sumda, and Chogdo Formations, dated at c. 55–50 Ma, via U–Pb zircon and/or biostratigraphy (op.cit.). The Jurutze Formation comprises black shales, siltstones, phyllites, and sandstones, which conformably grade upwards into the Sumda Formation containing nummulitic limestones, bioclastic wackestones, and sandstones, again conformably grading into the Chogdo Formation which comprises a fining-upwards sequence of conglomerates, gritstones, and siltstones. These formations were deposited within an arc-bound marine basin. Increasing amounts of sandstone and terrigeneous material present within the upper parts of the stratigraphy indicate the proximity of coastal environments and closing marine basins. The Nummulitic Limestone Formation marks the oldest fully marine sequence present in Ladakh.

The present outcrop width of the Indus Molasse Formations that overlie the Nummulitic Limestone Formation varies between 15 and 20 km. Some workers

Fig. 2.5 The Hemis Conglomerate Formation is some 1500 m thick around the type area of Hemis, 30 km SE of Leh. **a** Presents a closer view of medium-thickly bedded red coarse sandstones and conglomerates. **b** Gives an overview of the locality: the rocks are dipping to the south and thrusted towards the north. Note the variable bedding thicknesses from c. 1 m or less to 10 m and the variation in grain size of the red beds. **c** Presents a more vertical view of thrusted red beds, mostly thickly to medium bedded. Note the coarse nature of the beds beneath the Buddha statue

suggest that the Molasse is unconformable both on the Tar Group and the Ladakh Batholith (e.g. Searle et al. 1990), whilst others suggest a degree of conformity between the Molasse units and the underlying Tar Group (Henderson et al. 2010). The Eocene base of succession is dated, as indicated above, through biostratigraphy and U–Pb zircon dating from the Zanskar gorge area (Van Haver 1984 and Henderson et al. 2010) with other parts of the sequence dated at Upper-Eocene and Lower Oligocene from palm leaves (Van Haver 1984), and early Miocene (Brookfield and Andrews-Speed 1984a, b). Henderson et al. (2010) date the whole Molasse sequence through U–Pb zircon and Ar–Ar mica methods at between 51 and 24 Ma with one 6.1 Ma age for the uppermost stratigraphy. Zhou et al. (2020) track the unconformity between the Molasse and the Ladakh batholith over a strike length of 25 km in the Basgo-Tia area, c. 30 km WNW of Leh. This stratigraphic level is marked by pebble-boulder sized granitoid clasts within the molasse.

The lowermost Nurla Formation is stratigraphically immediately above the Nummulitic Limestone Formation. In the Zanskar Gorge area it is c. 700 m thick. The sequence comprises a sandstone: mudstone ratio of c. 60:40. Coarse sandstones are thickly bedded with individual beds some 15–20 m thick, interbedded with medium-thickly bedded siltstones, 8–20 m thick. Pebbly conglomerates can form the basal deposits of sandstone layers. There is an overall upwards-fining

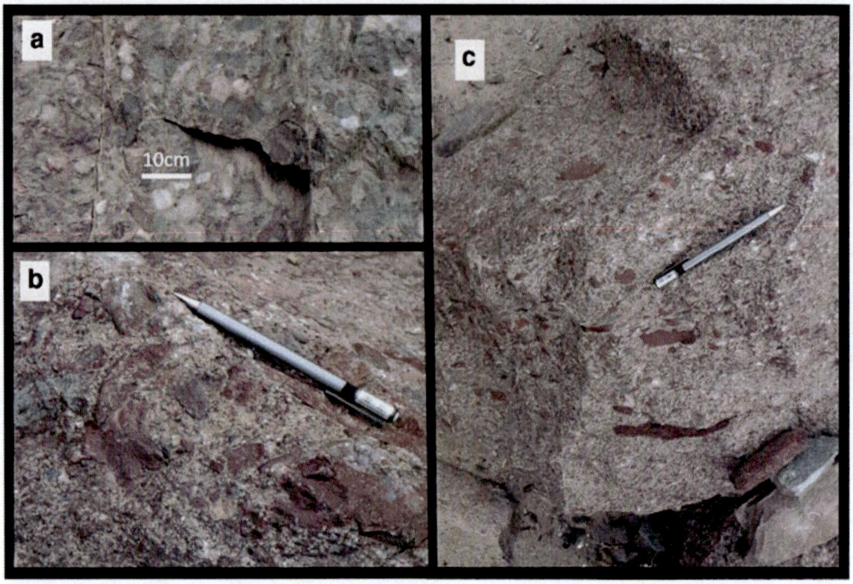

Fig. 2.6 Examples of outcrops from the Zanskar Gorge of the Choksti Conglomerate Formation. **a** Presents a clast supported facies with largely rounded to sub-rounded larger clasts. **b** Presents a high-large clast concentration conglomerate with a coarse sandstone-gritstone matrix. **c** is more matrix rich with elongate lenses of mudstone among the larger clasts. Note the heterolithic nature of the conglomerate, the strong red colour of many clasts and much of the matrix, and the extremely poor sorting. Pencil is 15 cm long

progression to the stratigraphy. Sedimentary structures include tabular, lensoid, cross-bedded and trough cross-bedding. Interbedded mudstones are red to dark grey and are normally-graded. Fossil leaf and wood is locally present. Clast supported conglomerates contain clasts up to 15 cm, with lithologies including pelites, micrites, chert, and granodiorite. Around Choksti tabular-bedded coarse-medium sandstones are interbedded with graded grey silt-mudstones. Palaeo-currents are to the south and southeast, with a minority to the north. Facies interpretation includes meandering braided river systems with pebbly channels, and floodplain over bank deposits. Some units have turbiditic characteristics. The Nurla Formation is dated at c. 51 Ma by U–Pb zircon methods, Henderson et al. (2010).

The Choksti Conglomerate overlies the Nurla Formation and is some 300 m thick. The lower 100 m comprises medium-thinly bedded sandstones with palaeocurrents to the NE and East. The upper 200 m comprises up to 8 m thick sequences of conglomerates. These are clast-supported, very poorly sorted, with a coarse sandstone matrix. Larger clasts are up to 30 cm in size, and are are rounded: smaller clasts are more angular. Clast lithologies include meta-sediments, volcanics, and granitoids. Clasts in the lower stratigraphy are dominated by granitoids. Facies interpretations include alluvial fans, deep-water conglomerate, sheet flood (mudstone interbeds) and a continental environment. The Choksti Formation is dated at c. 46 Ma by U–Pb zircon methods, Henderson et al. (2010).

The Hemis Conglomerate Formation overlies the Choksti Conglomerate Formation. This unit is up to 1500 m thick around Hemis, but only 150 m thick in the Zanskar Gorge, with an estimated Late Eocene-Early Oligocene age (Figs. 2.5 and 2.6). Some 20–50 m thick units can be followed for considerable distance along strike. Lowermost units contain granitoid boulders up to 3 m in size, with channels cutting 10 m into the substrate. Red mudstones interbed with conglomerates. Some clasts are well rounded, imbricated, and clast supported, with clast sizes up to 30 cm. The clasts are 90% granitoid with acid lavas and tuffs and meta-sediments also present. Interbedded coarse sandstones are trough cross-bedded. Burrows, rootlets, and wood fossils are locally present. Facies interpretations include high-energy fluvial channels, floodplain mud deposition, fan-delta, and local turbiditic affinities. There are no paleocurrent data from the Hemis area, but the equivalents In the Zanskar gorge are mostly north-directed.

The Uppermost, youngest stratigraphic unit is the 600 m thick Nimu Formation. This unit consists of fine-grained sandstones and siltstones, together with mudstones and local pebble-rich coarse bases to the sandstones. Sedimentary structures include loading structures, flute casts, ripples, and groove-casts. Freshwater bivalves are present along with burrowing structures. Palaeo-currents indicate N-NE and East directed to SE-SSW directed flows. The sequence is interpreted as a turbidite facies possibly within a delta-slope environment. Henderson et al. 2010 date the Lower Nimu stratigraphy at 24 Ma and the Uppermost Nimu stratigraphy at c. 6 Ma, using the Ar–Ar mica method.

2.3 Broad Palaeo-environments and Unroofing Evidence Recorded by the Indus Suture Rocks

The Indus Suture Rocks record two broad palaeo-environments. Stratigraphy within and beneath the Nummulitic Limestone Formation record sedimentation within a closing Tethys Basin culminating in depositional environments proximal to continental margins and a final marine incursion during the time of the Nummulitic Limestone (e.g. Searle et al. 1990; Sinclair and Jaffey 2001; Henderson et al. 2010, 2011). Above the Nummulitic Limestone Formation the rocks record a range of sub-environments within a gross intra-montane, intra-continental, post-continental collisional environment.

The general term 'molasse' refers to fluvial sedimentary processes within an eroding mountainous environment (Figs. 2.5, 2.6, 2.7, and 2.8). The overall picture (e.g. Searle et al. 1990) is one of a c. 200 km long by c. 60 km intra-montane, intra-continental basin, within which 2–3 km of sediment was deposited (Fig. 2.8). The sediment was derived from the erosion of surrounding mountains through the action of ice, wind, and liquid water, the transportation of this sediment via fluvial processes, and deposition within alluvial fan, fluvial channel, river overbank, lacustrine, and delta-turbiditic environments. The coarsest conglomerates contain clasts with long axes up to 3 m long, mostly 10–30 cm long (Figs. 2.6 and 2.7). Deposition of such coarse material was within proximal alluvial fan environments and large high-energy rivers, during times of high to very high flow (Figs. 2.6, 2.7, and 2.8). Sandstone, gritstone, and pebbly sandstone facies, with numerous cross bedding structures were largely deposited by medium–high energy rivers within braided channels and deltaic environments (Figs. 2.7 and 2.8). Other sandstone units with erosional bases, flute casts, load-structures, and cyclical coarse-fining upwards graded units were deposited by turbiditic fluvial processes. Mudstones and fine siltstones, many of which are red or red-green in colour, are interpreted as the products of quiet, low-energy overbank floodplain deposits and deposition within lakes. Fossil plants, rootlets, freshwater molluscs, burrow trails, and wood all evidence subaerial to lacustrine subaqueous environments (Fig. 2.8).

Fig. 2.7 Modern alluvial fan deposit in the Zanskar Gorge. This proximal deposit is strongly red coloured, very poorly sorted, and contains clasts that vary from mud to boulder size within a fan geometry (**a**). **b** Presents a well-sorted, strongly rippled and current bedded/laminated grey silt-fine sandstone with gravel lag deposit close to the letter B. **c** Presents an example of the very poorly sorted, heterolithic, densely packed breccia-conglomerate with angular-sub-angular clasts. **b** and **c** are spatially close together in the sequence being typical sub-facies evidencing a range of fluvial environments from high-energy water-rich fluvial processes to hyper-concentrated debris flows. These modern deposits present windows into the likely environments of parts of the Chokst i and Hemis Formations within the Indus Molasse

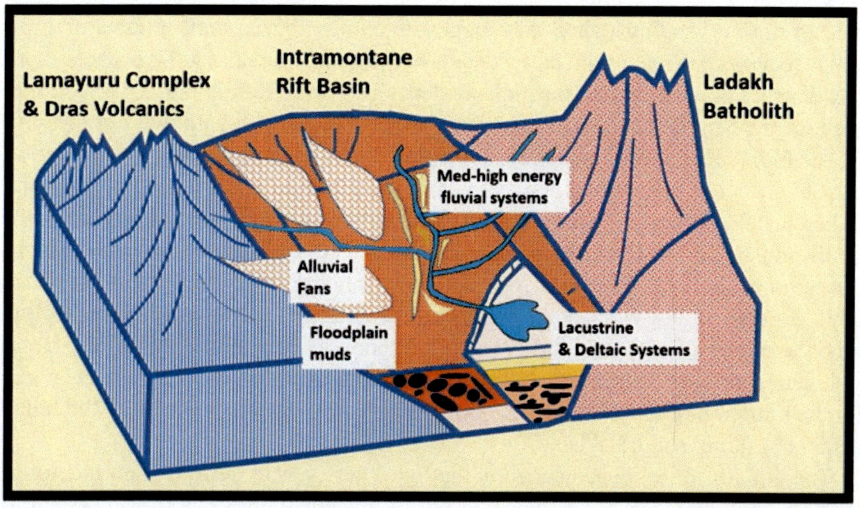

Fig. 2.8 Simplified diagram of the palaeo-environment of the Indus Molasse. Mountainous relief uplifts the Ladakh batholith, Dras Volcanics and Lamayuru Complex as well as terranes further afield. Intramontane rift valleys form and accumulate up to? 5 km of sediment from Eocene to Miocene times. Uplifted mountains generate abundant coarse sediment redistributed as alluvial fans, braided and meandering river deposits, and lacustrine-deltaic deposits, alongside mudstones within the overbank facies. The continental, highly oxidising environment produces a strong red colour in the rocks

The post-Nummulitic Limestone Indus Suture Rocks stratigraphy records a continuous-semi-continuous sequence of sedimentation from the Eocene to the Upper Oligocene, or < 50 to 51 Ma to as young as c. 6 Ma: some 46 Ma of sedimentation.

Recorded clasts within the Molasse units include Ladakh batholith granitoids, volcanic rocks, meta-sediments, cherts, pelites, and micrites. Palaeo-current directions are predominantly from the northern palaeo-terrane (these would contribute the Ladakh batholith and volcanic rock clasts) although a minority are from the southern palaeo-terranes (these would contribute the meta-sediment, limestone, chert clasts). Some workers (e.g. Searle et al. 1990) suggest that east–west river flows along strike within the molasse basin were also common. Clift et al. (2002) proposes that the absence of clasts from the High Himalaya, south of Zanskar, to the south is evidence for predominant uplift occurring only in the Ladakh batholith (and to a lesser extent in the southern Zanskar) region, and pre-dating the Main Central Thrust activity which formed the High Himalayan uplift period at c. 17–23 Ma. However, dates within the Indus Suture Rocks as young as 6 Ma (Henderson et al. 2010) suggest that the molasse records a much more protracted erosional and sedimentation period than Clift et al. (2002) anticipated: the principle of zero input of High Himalayan clasts remains cogent, suggesting the presence of topographic barriers at the time between the High Himalaya and the Indus Molasse basin. Sinclair and Jaffey (2001) present illite crystallinity data that suggest maximum burial temperatures of 155–180 °C equating to maximum burial depths of 3–5 km. Wu et al. (2007) propose three provenances of clasts within the molasse deposits: (1) the Karakoram-Lhasa terrane, north of the Shyok Suture; (2) the Trans-Himalayan-Ladakh batholith terrane; and (3) the Indian Plate. The Ladakh batholith-KLA terrane is by far the most important clast-contributor, with minor inputs from the Karakoram and later inputs from the Indian plate. Zhou et al. (2020) interpret an exclusive northerly directed provenance for the molasse in the Basgo area. Henderson et al. (2010) propose an ever-deepening dissection model with time, with sediment provenance predominantly from the north, until Miocene times, when some sediment had an Indian Plate provenance. At this time the modern Indus longitudinal river morphology was developing transporting sediment east–west. Sinclair and Jaffey (2001) suggest that an early Miocene age for the initiation of the Indus River, also evidenced by the development of the Indus delta in Makran, south Pakistan.

The bulk of the evidence suggests that most clasts were derived from the north, with contributions also from the south during Eocene–Oligocene times. Granitoid and volcanic clasts dominate Eocene–Oligocene molasse sequences. The Eocene aged molasse unconformity contains large granitoid boulders and deep channels, evidencing the rapid uplift and deep dissection of the Ladakh batholith at this time. Granitoid-rich conglomerates are common during Oligocene times (e.g. the Choksti and Hemis Conglomerates). In Miocene times sediment provenance was increasingly from both the north and south as the proto-Indus developed within a broadly east–west geography, with N–S and S–N flowing tributary rivers, as is the case today in Ladakh. Uplift, unroofing, and erosion of the Ladakh batholith was at its maximum during Eocene–Oligocene times. Some workers believe that sedimentation within the Indus Basin ceased or greatly diminished from Early Miocene times (e.g. Sinclair and Jaffey 2001).

2.4 Structure, Deformation, and Uplift of Ladakh, with References to Kohistan

The central study of this work is the structure and deformation of the Ladakh batholith. It is vital to set this within the context of the gross structure of Ladakh with reference also to Kohistan structure.

For Ladakh the most detailed structural studies have been undertaken by Searle and co-workers (Searle 1983, 1986, 1991, 2019; Searle et al. 1988, 1990, 1992, 1997, 1998, 1999; Searle and Treloar 2010). Searle et al. (1988) present a clear and synthetic overview of the structural evolution of the Ladakh region (Fig. 2.9). In this model the main structures and chronology are: (1) pre-collisional high pressure, low-temperature, blueschist facies metamorphism and related deformation in Jurassic–Cretaceous times within the KLA (Dras) island arc; (2) the south-westwards obduction of the Spontang ophiolite during pre-Eocene times (pre-50 Ma) related to the closure of Tethys Ocean; (3) Piggy-back style SW-directed thrusting, with later north-directed back-thrusting, and normal extensional faulting. The youngest rocks affected by this deformation phase are Palaeocene-early Eocene in age, and this deformation produced a series of foreland basins which filled with molasse; (4) Late Tertiary, Miocene-Pliocene, North-directed back-thrusting including deformation of the Indus Suture molasse rocks. Structural trends from SW to NE vary from NE-dipping, SW-directed in the southern region, vertical in the central region, and southerly-dipping, northwards-directed in the northern region. The central region forms a 'pop-up-structure' with divergent thrusts direction, to the south and north either side of a central vertical structure. Thrust stack-generated crustal thickening resulted in a thermal maximum and widespread Barrovian style metamorphism within the High Himalaya, south of Zanskar, in Oligocene–Miocene times. Thrusts are associated with folding of various styles (isoclinal, tight, upright) within less competent strata, and extensive shear zones. In south Zanskar there is a large shear zone with extensional fabrics that separates high grade metamorphic rocks of the High Himalaya, from the low-grade metamorphic rocks of the Zanskar area. This structure probably accommodated 25–30 km of erosion. Searle (1986) notes additional observations to the above with respect to Ladakh structure: (1) thrusting took place in the region from c. 70 Ma to the Pliocene; (2) the South to Southwest directed Zanskar/southern Ladakh thrusts shortened the crust by c. 126 km; (3) thrusting and related folding are deep seated structures within the Himalayas reaching mid-lower crustal levels, even to the base of the crust; (4) The Zanskar shear structure separates sillimanite-kyanite migmatites to the south from lower Greenschist rocks to the north, linked to orogenic collapse; (5) the Tso Morari thrust in Ladakh extends from the lower-mid crust to the surface with throws of 5–10 km; (6) the S-SW thrust dominated deformation stage extended from c. 45–25 Ma and involved thick sections of Palaeocene–Eocene crust as well as listric normal faults and the Zanskar Shear Zone (with 2–7 km of fault-throw) and (7) the north directed back-thrusts and crustal pop up are estimated to have formed from 15 Ma.

For Kohistan Coward and co-workers have presented some of the most detailed and terrane-wide structural analyses (Coward et al. 1982a, b, 1986, 1987) together with a modern synthesis published by Searle and Hacker (2019): see Fig. 2.9 for a simplified structural cross-sections. The Kohistan structural model presents: (1) two collisional events with cleavage and folding developed at c. 90–100 Ma close to the Northern-Shyok Suture, and similar deformation in deeper crustal Kohistan close to the Indus Suture; and (2) Tehys ocean closure during the Eocene and obduction of Kohistan over the Indian continental crust. Post Eocene thrusting is towards the SSE with complex refolding and thrust patterns close to the Hazara and Nanga Parbat Syntaxes. The gross structural model of Kohistan as proposed by Coward and co-workers have similar characteristics to elements of Ladakh structure in the sense that: (1) south-directed thrust structures predominate to the south; (2) there is a central zone of upright structures, folds and thrusts (the Chilas antiform and Jaglot synform in Kohistan), and (3) north-directed structures to the north. This structural style in Kohistan was mainly formed pre-90 Ma, due to the collision between Kohistan and

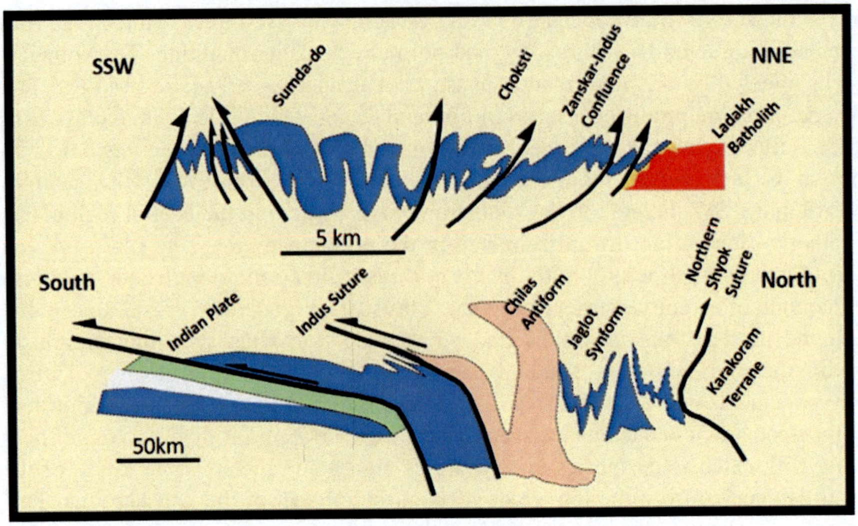

Fig. 2.9 Simplified Structural Cross Sections across Kohistan Ladakh. Upper section is simplified from Sinclair and Jaffey (2001) (and earlier Searle et al. 1988) acknowledgement Geological Society of London, from the Zanskar valley/Gorge section. Whilst early structures are S-SW directed, these become upright structures in the central parts with north-directed back-thrusts dominating this part of the section. The Indus Molasse is unconformable on and thrust-over the Ladakh batholith. The lower section is simplified from Coward et al. (1986) acknowledgement Geological Society of London, and displays a whole-KLA terrane structural cross section. The southern part of Kohistan and the Indian Plate is strongly affected by south-directed thrusts. The central part of Kohistan around Chilas-Jaglot is folded and has a strong vertical-upright-overturned character. The Kohistan batholith occupies the space north of Jaglot within an antiformal structure. North-directed thrusts and shears dominate the part of Kohistan affected by the Northern-Shyok Suture structures

Eurasia/Karakoram, and from 80 Ma, and particularly post-Indian collision, south-directed thrusting became dominant, with crustal-scale shear zones affected the lower to mid crustal levels in southern Kohistan.

It is instructive to examine studies of the Nanga Parbat Syntaxis (NPS) which bounds much of western Ladakh and eastern Kohistan (Fig. 2.10). The NPS forms an Indian Plate promontory almost cutting the Kohistan–Ladakh arc in two (Fig. 1.1). Syntaxes are quasi-linear-arcuate major structures, cored by older rocks, and separated from surrounding younger rocks by thrusts, faults, shears, and folds. The Himalayas contain several syntaxial structures, of which the Nanga Parbat Syntaxis is the largest. Butler (2019) reviewed a range of research that had been undertaken over a 40 year period with respect to Himalayan Syntaxes and a summary of Butler's main findings are presented below. The NPS has several characteristics which make it a critical structure in the NW Himalayan region. These characteristics include: (1) the production of the > 8000 m high Nanga Parbat-Haramosh massif; (2) the presence of ancient Indian Plate (Proterozoic to Phanerozoic) metamorphic rocks juxtaposed and surrounded by younger terranes such as the KLA; (3) young cooling ages indicating exceptionally rapid recent uplift and exhumation rates; (4) the presence of a range of thrust and fault structures that separate the NPS from the KLA; (5) the existence of glaciers and deeply incised rivers including the River Indus which rapidly erode material from the central massif and transport sediment away from the region.

Most Himalayan syntaxes are young features, forming within the past 10 Ma. The NPS largely comprises high-grade migmatitic sillimanite-cordierite-K-feldspar gneisses flanked by biotite schists, psammites, and Phanerozoic biotite granites. Peak metamorphism of eclogites is estimated at 750 °C and 2.5 GPa (25 kb), overprinted by amphibolitic metamorphism which peaked at 650 °C and 0.7 GPa (7 kb) (Butler 2019, and references therein). The gross structure of the Nanga Parbat syntaxis is antiformal to double-antiformal and domal. Various tectonic mechanisms have been proposed for the formation of the NPS including folding, thrusting, pop-up divergent direction thrusting, and gravitational ductile flow-spreading. It is currently thought that tectonics alone may have been the dominant process during the earlier stages of syntax formation, with a feedback mechanism linked to rapid uplift, exhumation, denudation, and erosion becoming at least as dominant as tectonism during later stages of syntaxis evolution.

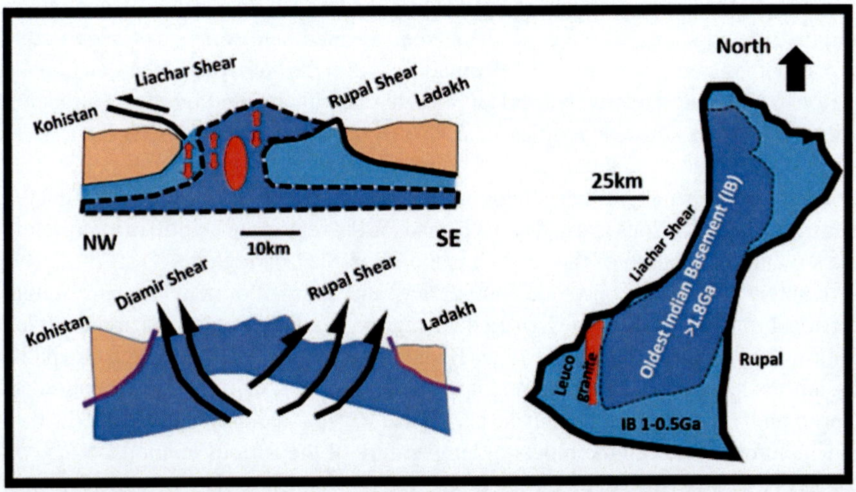

Fig. 2.10 Simplified Geological Map and cross sections of the Nanga Parbat Syntaxis after Butler (2019) and Schneider et al. (1999) Acknowledgement Geological Society of London. The Nanga Parbat syntaxis (NPS) is oriented NNE-SSW and separates the Indian Plate from the Kohistan–Ladakh arc, almost splitting the KLA in two separate parts. The NPS is cored by Proterozoic migmatitic gneisses, surrounded by youngest Proterozoic-Phanerozoic schists into which are intruded extremely young (c. 0.7 Ma) leucogranites formed through decompressional melting. The rapid rise of the NPS is explained through thrust-tectonic-erosion plus denudation models and/or tectonic-denudation and gravity sliding models. The lower left cross section explains the structure of the antiformal syntaxis predominantly in terms of thrusts and pop-up structures, whilst the upper left cross section acknowledges the role of thrusts and shears but emphasises simple shear and stretching fabrics produce through gravity sliding and gravitational flow

The most sophisticated tectonic model is possibly the tectonic aneurysm model which involves pop-up structures and antiforms/domes being largely formed from divergent thrust directions, with intense denudation and erosion continually assisting rapid uplift and strain-softening rock rheology (Schneider et al. 1999; Zeitler and Chamberlain 1991). Alternative models suggest that gravity spreading is the main tectonic process, alongside thrusting, with ductile, vertical, pure-shear structures being predominant in much of the structure. A period of crustal thickening and warming, reducing crustal rheological strength is considered a probable pre-requisite for syntax formation, with crustal warming taking tens of millions of years (Butler 2019). Nanga Parbat has risen remarkably quickly producing incision rates of 4–6 m/ka for the period 27–65,000 years ago and 10–14 m/ka from 7000 years ago, with present day uplift rates as high as 6 mm/year (Zeitler 1985; Butler 2019 and references therein). The rapid rise of Nanga Parbat has led to decompressional melting, and the production of leucogranites as young as 0.7 Ma (George et al. 1993). In-situ melts and melt surrounding grains may be one mechanism for reducing the rheological strength of the syntaxis, at least locally. Geothermal gradients may reach as high as 120 °C per km with the ductile–brittle transition zone reached at depths of 5–6 km and temperatures that may vary from 350 °C to 425–600 °C. Butler (2019) argues that thrusting tectonics may dominate during earlier stages, localising deformation within relatively colder crust, with gravity spreading producing more of a ductile-continuum and distributed strain, as the crust thickened, warmed, and softened. Nanga Parbat assists with the understanding of deformation within the batholith of the Leh Ladakh region as (1) being a relatively local large-scale structure it has a significant structural influence within adjacent regions; and (2) it presents a series of structural styles within crystalline rock and quantified boundary conditions in terms of temperature, pressure, geothermal gradient, rheological controls, and defining the brittle-ductile transition zone. Lessons learned in the Nanga Parbat region are helpful in interpreting the structural evolution of the Ladakh batholith in the Leh region.

Kirstein et al. (2009) provide an overview and analysis of new low-temperature thermochronology data in the Leh-Ladakh region and northwards towards the Shyok Suture Zone and Karakoram terrane. Radiometric dating methods, such as apatite fission tracks (AFT) and zircon and apatite helium ages, using U–Th/He ratio data (shorthand ZrHe for zircon analysis and ApHe for apatite analysis) provide the date at which rocks reached the blocking temperature of the various methods: c. 200 °C for ZrHe, 120 °C for AFT and 55–80 °C for ApHe. These data provide evidence for the age at which the crust reached various crustal isotherms (200 °C, 120 °C and 80–55 °C, respectively). These isotherms correspond to shallow-medium crustal levels depending on local geothermal gradients (Gautheron et al. 2022). The average geothermal gradient of 25–30 °C /km would mean that crustal temperatures of 200, 120, and 55 °C, would be met at depths of c. 7–8 km, 4–5 km and c. 2 km respectively although these depths would vary according to the local geothermal gradient quantum (Gautheron et al. 2022). The interpretation of data in terms of crustal depths and crustal uplift/subsidence rates is model-driven to some extent, although realistic contextual analysis (e.g. in this case uplift within a higher-geothermal gradient,

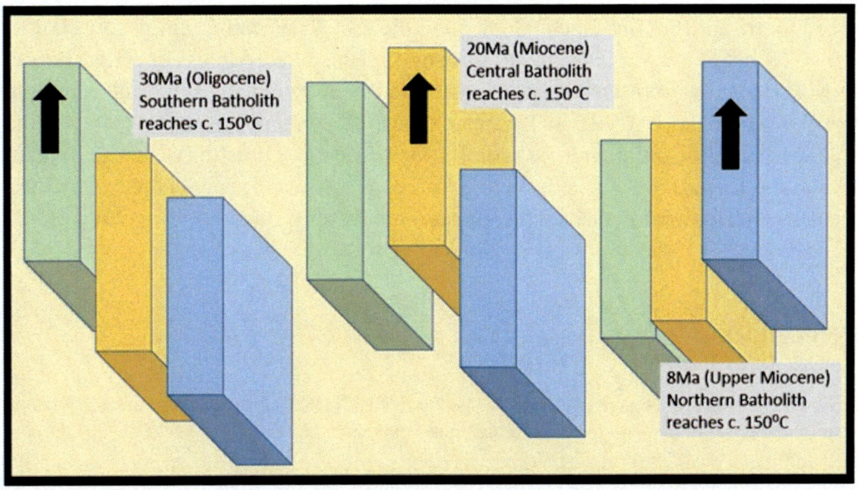

Fig. 2.11 Time-uplift diagram for the Ladakh batholith using low temperature data of Kirstein et al. (2009) and other workers cited in Kirstein et al. (2009). The three block diagrams show the approximate situations of the southern, central, and northern parts of the Ladakh batholith within the Leh-Ladakh area as they pass through the c. 150 °C isotherm. In simple terms the low temperature age data young northwards. Strong denudation occurred from 15 to 6 Ma leading to rapid uplift in the northern part of Ladakh, linked to tectonics in the northern Karakoram terrane. The Molasse begins to thrust northwards from c. 20 Ma and blankets part of the southern batholith, protecting it from denudation

active continental collision zone setting, involving migmatitic metamorphism and leucogranite crustal melt genesis) reduces uncertainty.

In their review of low-temperature geothermometry data from the Leh-Ladakh and broader Ladakh region Kirstein et al. (2009) observe that ages vary from 44–4.8 Ma, mostly 44–10 Ma, indicating a Late Eocene-Miocene crustal cooling pattern, with data from the Khardung La region indicating rapid uplift/exhumation rates of 0.1 km/Ma at 22 Ma. Kirstein et al. (2009) then present new data from a south-north traverse starting at the Indus River and ending in the Shyok area. Data in general show a younging of age data moving northwards from the Indus River. The southern batholith area yields ZHe ages of 25–33 Ma and AFT ages of 18–21 Ma; the central batholith area yields ZHe ages of 22–24 Ma and AFT ages of 35–30 Ma; and the northern batholith area yields ZHe ages of 17–13 Ma and AFT ages of 5.7–6.9 Ma. Cooling-time modelling curves suggest that the southern batholith crustal area achieved temperatures of 200 and 40 °C at 45–30 Ma and 25–10 Ma, the central batholith at 25 Ma and 12 Ma and the northern region of Ladakh at 10 Ma and 4 Ma respectively. The area experienced high exhumation/uplift rates of > 0.4 km/Ma during maximum periods of uplift. The overall crustal model (Fig. 2.11) proposes: (1) 49–30 Ma: a northern tilting and uplift in the Khardung area associated also with crustal doming. The Indus River area cooled below 150 °C at 30 Ma; (2) 30–15 Ma: southward-tilting of the batholith during Oligocene–Miocene times, in part linked to the uplift of the Tibetan Plateau to the east. The central region cools below 150 °C at c. 22–18 Ma. Northward thrusting of the Indus Molasse occurs from c. 20 Ma; (3) 15 Ma-present: strong erosion and denudation in the northern region, slower uplift from c. 6 Ma in Pliocene times. The molasse blankets parts of the southern batholith, protecting the batholith from erosion. Thrusting of the Karakoram terrane southwards over Ladakh occurs, with both north- and south-directed thrusting occurring in southern Ladakh. The northern region cooled below 150 °C at c. 8 Ma.

References

Ahmad T, Thakur VC, Islam R, Khanna PP, Mukherjee PK (1998) Geochemistry and geodynamic implications of magmatic rocks from the trans-Himalayan arc. Geochem J 32:383–404. https://doi.org/10.2343/geochemj.32.383

Andjić G, Zhou R, Jonell TN, Aitchison JC (2022) A single Dras-Kohistan-Ladakh arc revealed by volcaniclastic records. Geochem Geophys Geosyst 23:e2021GC010042

Bard JP (1983) Metamorphism of an obducted island arc: example of the Kohistan sequence (Pakistan) in the Himalayan collided range. Earth Planet Sci Lett 65:133–144

Bard JP, Muluski H, Matte P, Proust F (1980) The Kohistan sequence: crust and mantle of an obducted island arc. In: Tahirkheli RAK, Jan MQ, Majid M (eds) Proceedings of the international commission on geodynamics, special issue, vol 13. Geological Bulletin University Peshawar, pp 87–94

Bignold SM, Treloar PJ (2003) Northward subduction of the Indian Plate beneath the Kohistan island arc, Pakistan Himalaya: new evidence from isotopic data. J Geol Soc Lond 160:377–384

Bignold SM, Treloar PJ, Petford N (2006) Changing sources of magma generation beneath intra-oceanic island arcs: an insight from the juvenile Kohistan island arc, Pakistan Himalaya. Chem Geol 233:46–74

Brookfield ME, Andrews-Speed CP (1984a) Sedimentology, petrography and tectonic significance of the shelf, flysch and molasse elastic deposits across the Indus suture zone, Ladakh, NW India. Sed Geol 40:249–286

Brookfield ME, Andrews-Speed CP (1984b) Sedimentation in a high-altitude intermontane basin-the Wakka Chumolasse (mid-Tertiary, northwestern India). Bull Indian Geol Assoc 17:175–193

Butler RWH (2019) Tectonic evolution of the Himalayan syntaxes: the view from Nanga Parbat. In: Treloar PJ, Searle MP (eds) Himalayan tectonics: a modern synthesis, vol 483. Geological Society, London, Special Publications, pp 215–254. https://doi.org/10.1144/SP483.7

Clift PD, Hannigan R, Blusztajn J, Draut AE (2002) Geochemical evolution of the Dras-Kohistan Arc during collision with Eurasia: evidence from the Ladakh Himalaya, India. Islands Arc 11:255–273

Clift PD, Pavlis T, DeBarr SM, Draut AE, Rioux M, Kelemen PB (2005) Subduction erosion of the Jurassic Talkeetna-Bonanza arc and the Mesozoic accretionary tectonics of western North America. Geology 33:881–884

Coward MP, Jan MQ, Rex D, Tarney J, Thirlwall M, Windley BF (1982a) Geotectonic framework of the Himalaya of N Pakistan. J Geol Soc Lond 139:299–308

Coward MP, Jan MQ, Rex D, Tarney J, Thirlwall M, Windley BF (1982b) Structural evolution of a crustal section in the western Himalaya. Nature 295:22–24

Coward MP, Butler RWH, Khan MA, Knipe RJ (1987) The tectonic history and its implications for Himalayan structure. J Geol Soc London 144:377–391

Coward MP, Windley BF, Broughton RD, Luff IW, Petterson MG, Pudsey CJ, Rex DC, Khan MA (1986) Collision tectonics in NW Himalayas. In: Coward MP, Ries AC (eds) Collision tectonics, vol 19. Geological Society of London Special Publication, pp 203–219

Desio A (1959) Cretaceous beds between Karakorum and Hindu Kush ranges (Central Asia). Rivista Italiana Di Paleontologia Strat 65:221–229

Desio A (1964) Geological tentative map of the western Karakorum, 1:500,000. Istituto di Geologia, Milan

Gansser A (1964) Geology of the Himalayas. Wiley Interscience, London, p 289

Gansser A (1979) Map of ophiolitic belts of Himalayan and Tibetan region. In: International geological correlation programme international atlas of ophiolites, 1:500 000. Geological Society of America, Boulder, Colorado

Gansser A (1981) The geodynamic history of the Himalaya. In: Gupta HK, Delany FM (eds) Zagros, Hindu Kush, Himalaya, geodynamic evolution, vol 3. American Geophysical Union, Geodynamic Series, pp 111–121

Garzanti E, van Haver T (1988) The Indus clastics: forearc basin sedimentation in the Ladakh Himalaya (India). Sed Geol 59:237–249

Gautheron C, Hueck M, Ternois S, Heller B, Schwartz S, Sarda P, Tassan-Got L (2022) Investigating the shallow to mid-depth (>100–300°C) continental crust evolution with (U–Th)/He thermochronology: a review. Minerals 12:563. https://doi.org/10.3390/min12050563

George MT, Harris NBW, Butler RWH (1993) The tectonic implications of contrasting granite-magmatism between the Kohistan island arc and Nanga Parbat-Haramosh massif, Pakistan Himalaya. In: Treloar PJ, Searle MP (eds) Himalayan tectonics, vol 74. Geological Society, London, Special Publication, pp 173–191

Hayden HH (1914) Notes on the geology of Chitral and Gilgit and the Pamirs. Rec Geol Surv India 45:271–335

Henderson AL, Najman Y, Parrish R, BouDagher-Fadel M, Barford D, Garzanti E, Andò S (2010) Geology of the Cenozoic Indus Basin sedimentary rocks: paleoenvironmental interpretation of sedimentation from the western Himalaya during the early phases of India-Eurasia collision. Tectonics 29:TC6015. https://doi.org/10.1029/2009TC002651

Henderson AL, Najman Y, Parrish R, Mark DF, Foster GL (2011) Constraints to the timing of India-Eurasia collision; a re-evaluation of evidence from the Indus Basin sedimentary rocks of the Indus-Tsangpo Suture Zone, Ladakh, India. Earth Sci Rev 106(3–4):265–292. https://doi.org/10.1016/j.earscirev.2011.02.006

Honegger K, Dietrich V, Frank W, Gansser A, Thoni M, Trommsdorff V (1982) Magmatism and Metamorphism in the Ladakh Himalayas (the Indus-Tsangpo Suture Zone). Earth Planet Sci Lett 60:253–292

Ivanac JF, Traves DM, King D (1956) The geology of the northwest portion of the Gilgit agency. Rec Geol Surv Pakistan 8(2):3–26

Jadoon UK, Lin D, Baral U, Qasim M (2019) Early cretaceous to Eocene magmatic records in Ladakh arc: constraints from U–Pb ages of Deosai volcanics, northern Pakistan. Geo J 1–14. https://doi.org/10.1002/gj.3730

Jagoutz OE (2009) Construction of the granitoid crust of an island arc. Part 1: geochronological and geochemical constraints from plutonic Kohistan (NW Pakistan). Contrib Miner Petrol 158:739–755

Jagoutz OE (2010) Construction of the granitoid crust of an island arc. Part 2: a quantitative petrogenetic model. Contrib Miner Petrol 160:359–381

Jagoutz O (2014) Arc crustal differentiation mechanisms. Earth Planet Sci Lett 396:267–277. https://doi.org/10.1016/j.epsl.2014.03.060

Jagoutz O, Kelemen PB (2015) Role of arc processes in the formation of continental crust. Ann Rev Earth Sci 43:363–404

Jagoutz O, Schmidt MW (2012) The formation and bulk composition of modern juvenile crust: the Kohistan arc. Chem Geol 298–299:79–96

Jagoutz O, Müntener O, Schmidt MW, Burg J-P (2011) The roles of flux- and decompression melting and their respective fractionation lines for continental crust formation: evidence from the Kohistan arc. Earth Planet Sci Lett 303:25–36

Jagoutz O, Bouilhol P, Schaltagger P, Muntner O (2019) The isotopic evolution of the Kohistan—Ladakh arc from subduction initiation to continent continent collision. In: Treloar PJ, Searle MP (eds) Himalayan tectonics: a modern synthesis, vol 483. Geological Society, London, Special Publications. https://doi.org/10.1144/SP483.7

Jain AK, Lal N, Sulemani B, Awasthi AK, Singh S, Kumar R, Kumar D (2009) Detrital-zircon fission-track ages from the Lower Cenozoic sediments, NW Himalayan foreland basin: Clues for exhumation and denudation of the Himalaya during the India-Asia collision. GSA Bull 121(3/4):519–535. https://doi.org/10.1130/B26304.1

Khan MA, Coward MP (1991) Entrapment of an intra-oceanic island arc in collision tectonics: A review of the structural history of the Kohistan arc complex, NW Himalayas. In: Sharma KK (ed) Geology and geodynamic evolution of the Himalayan collision Zone Part I. Physics and chemistry of the earth, vol 17, pp 1–18

Khan MA, Stern RJ, Gribble RF, Windley BF (1997) Geochemical and isotopic constraints on subduction polarity, magma sources, and paleogeography of the Kohistan intra-oceanic arc, northern Pakistani Himalayas. J Geol Soc Lond 154:935–946

Khan SD, Walker DJ, Hall SA, Burke KC, Shah MT, Stockli L (2009) Did the Kohistan-Ladakh island arc collide first with India? Geol Soc Am Bull 121:366–384

Kirstein LA, Foeken JPT, Van der Beek P, Stuart FM, Phillips RJ (2009) Cenozoic unroofing history of the Ladakh Batholith, western Himalaya. J Geol Soc Lond 166:667–678. https://doi.org/10.1144/0016-76492008-107

Kumar S (2008) Magnetic susceptibility mapping of Ladakh granitoids, northwest higher Himalaya: implication to redox series of felsic magmatismin the subduction environments. Mem Geol Soc India 72:83–102

Kumar S (2010a) Mafic to hybrid microgranular enclaves in the Ladakh batholith, northwest Himalaya: implications on calc-alkaline magma chamber processes. J Geol Soc India 76:5–25. https://doi.org/10.1007/s12594-010-0080-2

Kumar S (2010b) Magnetite and ilmenite series of granitoids of Ladakh Batholith, Northwest Indian Himalaya: implications on redox conditions of calc-alkaline felsic magma chambers. Curr Sci 99:1260–1264. https://doi.org/10.2307/24068524

Kumar S (2020) Schedule of mafic to hybrid magma injections into crystallizing felsic magma chambers and resultant geometry of enclaves in granites: new field and petrographic observations from Ladakh Batholith, trans-Himalaya, India. Front Earth Sci 8:551097. https://doi.org/10.3389/feart.2020.551097

Kumar S, Singh B (2008a) Mineralogy and Geochemistry of mafic to hybrid microgranular enclaves and felsic host of Ladakh batholith, Northwest Himalaya: evidence of multistage complex magmatic processes. Himalayan J Earth Sci 5:130–131. https://doi.org/10.3126/hjs.v5i7.1315

Kumar S, Singh B (2008b) Mineralogy and Geochemistry of mafic to hybridmicrogranular enclaves and felsic host of Ladakh batholith, Northwest Himalaya: Evidence of multistage complex magmatic processes. Himalayan J Sci 5:130–131. https://doi.org/10.3126/hjs.v5i7.1315

Kumar S, Singh B, Wu FY, Ji WQ (2012) Timing of Asia-India collision evident from U-Pb–Lu–Hf isotopes of zircons from granitoids and enclaves of Ladakh Batholiths, NW Himalaya. J Nepal Geol Soc 45:164–165

Kumar S, Bora S, Sharma U (2016) Geological appraisal of Ladakh and Tirit Granitoids in the Indus-Shyok suture zones of Northwest Himalaya, India. J Geol Soc India 87:737–746

Martin CR, Jagoutz O, Upadhyay R, Royden LH, Eddy MP, Bailey E, Nichols CIO, Weiss BP (2020) Paleocene latitude of the Kohistan-Ladakh arc indicates multistage India-Eurasia collision. Proc Natl Acad Sci 117(47):29487–29494. https://doi.org/10.1073/pnas.2009039117

Petterson MG, Treloar PJ (2004) Volcanostratigraphy of arc volcanic sequences in the Kohistan arc, north Pakistan: volcanism within island arc, back-arc basin, and intra-continental tectonic settings. J Volcanol Geoth Res 130:147–178

Petterson MG, Windley BF (1985) Rb-Sr dating of the Kohistan arc-batholith in the trans-Himalaya of north Pakistan, and tectonic implications. Earth Planet Sci Lett 74:45–57

Petterson MG, Windley BF (1986) Petrological and geochemical evolution of the Kohistan arc-batholith, Gilgit, N. Pakistan. Geol Bull Univ Peshawar 19:121–149

Petterson MG, Windley BF (1991) Changing source regions of magmas and crustal growth in the trans-Himalayas: evidence from the Chalt volcanics and Kohistan batholith, Kohistan, northern Pakistan. Earth Planet Sci Lett 102:326–341

Petterson MG, Windley BF (1992) Field relations, geochemistry and petrogenesis of the Cretaceous basaltic Jutal dykes, Kohistan, Northern Pakistan. J Geol Soc Lond 149:107–114

Petterson MG, Crawford MB, Windley BF (1993) Petrogenetic implications of neodynamism isotope data from the Kohistan batholith, North Pakistan. J Geol Soc Lond 150:125–129

Petterson MG (2010) Review of the geology and tectonics of the Kohistan island arc, north Pakistan. In: Kusky TM, Zhai M-G, Xiao W (eds) The evolving continents: understanding processes of continental growth, vol 338. Geological Society, London, Special Publications, pp 287–327

Petterson MG (2019) The plutonic crust of Kohistan and volcanic crust of Kohistan-Ladakh, North Pakistan/India: lessons learned for deep and shallow arc processes. In: Treloar PJ, Searle MP (eds) Himalayan tectonics: a modern synthesis, vol 483. Geological Society, London, Special Publications. https://doi.org/10.1144/SP483.4

Petterson MG, Windley BF, Luff IW (1991a) The Chalt volcanics, Kohistan, N Pakistan: high-Mg tholeiitic and low Mg calc-alkaline volcanism in a Cretaceous island arc. In: Sharma KK (ed) Geology and geodynamic evolution of the Himalayan collision zone. Physics and chemistry of the earth, vol 17, pp 19–30

Petterson MG, Windley BF, Sullivan M (1991b) A petrological, chronological, structural and geochemical review of Kohistan batholith and its relationship to regional tectonics. In: Sharma KK (ed) Geology and geodynamic evolution of the Himalayan collision zone part I. Physics and chemistry of the earth, vol 17, pp 1–18

Pudsey CJ (1986) The Northern suture, Pakistan: margin of a Cretaceous island arc. Geol Mag 123:405–423

Pudsey CJ, Coward MP, Luff IW, Shackleton RM, Windley BF, Jan MQ (1985a) Collision zone between the Kohistan arc and the Asian plate in N W Pakistan. Trans R Soc Edinburgh 76:463–479

Pudsey CJ, Schroider R, Skelton PW (1985b) Cretaceous (Aptian/Albian) age for island arc volcanics, Kohistan, N. Pakistan. In: Gupta VJ (ed) Contributions to the Himalayan Geology, vol 3. Hindustan Publishing Corporation, Delhi, pp 150–168

Pudsey CJ, Schroeder R, Skelton PW (1986) Cretaceous (Aptian-Albian) age for island arc volcanics, Kohistan, N. Pakistan. In: Gupta VJ (ed) Recent researches in geology. Palaeontology: stratigraphy and structure of Western Himalayas. Geology of Western Himalayas. Hindustan Publishing Company, Delhi

Ravikant V, Fu-Yuan W, Qiang-Wei J (2009) Zircon U-Pb and Hf isotopic constraints on petrogenesis of the Cretaceous-Tertiary granites in eastern Karakoram and Ladakh, India. Lithos 110:153–166. https://doi.org/10.1016/j.lithos.2008.12.013

Reuber I (1986) Geometry of accretion and oceanic thrusting of the Spontang ophiolite, Ladakh Himalaya. Nature 321:592–596

Reuber I (1989) The Dras arc: two successive volcanic events on eroded ocean crust. Tectonophysics 161:93–106

Saktura WM, Buckman S, Nutman AP, Bennett VC (2020) Late Jurassic Changmar Complex from the Shyok ophiolite, NW Himalaya: a Prelude to the Ladakh Arc. Geol Mag. https://doi.org/10.1017/S0016756820000400

Schaltegger U, Zeilinger G, Frank M, Burg JP (2002) Multiple mantle sources during island arc magmatism: U-Pb and Hf isotopic evidence from the Kohistan arc complex, Pakistan. Terra Nova 14:461–468

Schneider DA, Edwards MA, Kidd WSF, Khan MA, Seeber L, Zeitler PK (1999) Tectonics of Nanga Parbat, western Himalaya: synkinematic pluton-ism within doubly vergent shear zones of a crustal-scalepop-up structure. Geology 27:999–1002

Searle MP (1991) Geology and Tectonics of the Karakoram Mountains. John Wiley & Sons, 358 pp

Searle MP (1983) Stratigraphy, structure and evolution of the Tibetan-Tethys zone in Zanskar and the Indus suture zone in the Ladakh Himalaya. Trans R Soc Edinburgh Earth Sci 73:205–219. https://doi.org/10.1017/S0263593300009688

Searle MP (1986) Structural evolution and sequence of thrusting in the High Himalayan, Tibetan-Tethys and Indus suture zones of Zanskar and Ladakh, Western Himalaya. J Struc Geol 8:923–936. https://doi.org/10.1016/0191-8141(86)90037-4

Searle MP, Cooper DJW, Rex AJ (1988) Collision tectonics of the Ladakh-Zanskar Himalaya. In: Shackleton RM, Dewey JF, Windley BF (eds) Tectonic evolution of the Himalayas and Tibet. R. Soc, London, pp 117–150

Searle MP, Hacker BR (2019) Structural and metamorphic evolution of the Karakoram and Pamir following India–Kohistan–Asia collision Vol 483. Geological Society of London Special Publications. pp 555–582. https://doi.org/10.1144/SP483.6

Searle M, Pickering K, Cooper DJW (1990) Restoration and evolution of the intermontane lndus molasse basin, Ladakh Himalaya, India. Tectonophysics 174:301–314. https://doi.org/10.1016/0040-1951(90)90327-5

Searle MP, Waters D, Rex D, Wilson R (1992) Pressure, temperature and time constraints on Himalayan metamorphism from eastern Kashmir and western Zanskar. J Geol Soc Lond 149:753–773. https://doi.org/10.1144/gsjgs.149.5.0753

Searle M et al (1997) Structure of the north Indian continental margin in the Ladakh-Zanskar Himalayas: implications for the timing of obduction of the Spontang ophiolite, India-Asia collision and deformation events in the Himalaya. Geol Mag 134:297–316. https://doi.org/10.1017/S0016756897006857

Searle MP, Khan MA, Fraser JE, Gough SJ, Jan MQ (1999) The tectonic evolution of the Kohistan-Karakoram collision belt along the Karkoram highway transect, North Pakistan. Tectonics 18(6):929–949

Searle MP, Treloar PJ (2010) Was the Late Cretaceous–Paleocene obduction of ophiolite complexes the primary cause of crustal thickening and regional metamorphism in the Pakistan Himalaya? In: Kusky TM, Zhai M-G, Xiao W (eds) The evolving continents: understanding processes of continental growth, vol 338. Geological Society, London, Special Publications, pp 345–359

Searle MP, Weinberg RF, Dunlap WJ (1998) Transpressional tectonics along the Karakoram fault zone, northern Ladakh: constraints on Tibetan extrusion. In: Holdsworth RE, Strachan RA, Dewey JE (eds) Continental transpressional and transtensional tectonics, vol 135. Geological Society, London, Special Publications, pp 307–326

Searle MP (2019) Timing of subduction initiation, arc formation, ophiolite obduction and India–Asia collision in the Himalaya. In: Treloar PJ, Searle MP (eds) Himalayan tectonics: a modern synthesis, vol 483. Geological Society, London, Special Publications, pp 19–37. https://doi.org/10.1144/SP483.8

Sinclair HD, Jaffey N (2001) Sedimentology of the Indus Group, Ladakh, northern India implications for the timing of initiation of the palaeo-Indus River. J Geol Soc Lond 158:151–162. https://doi.org/10.1144/jgs.158.1.151

Tahirkheli RAK, Jan MQ (1979) A preliminary geological map of Kohistan and the adjoining areas, N Pakistan, vol 11. Geological Bulletin, Peshawar University

Tahirkheli RAK (1979) Geology of Kohistan and adjoining Eurasian and Indo-Pakistan continents, Pakistan. In: Tahirkheli RAK, Jan MQ (eds) Geology of Kohistan, vol 11. Special Issue, Geological Bulletin, University of Peshawar, pp 1–30

Thakur VC (1981) Regional framework and geodynamic evolution of the Indus-Tsangpo suture zone in the Ladakh Himalayas. Trans R Soc Edinburgh Earth Sci 72:89–97

Thanh NX, Itaya T, Ahmad T, Kojima S, Ohtani Y, Ehiro M (2010) Mineral chemistry and K-Ar ages of plutons across the Karakoram fault in the Shyok-Nubra confluence of northern Ladakh Himalaya, India. Gondwana Res 17:180–188

Treloar PJ, Rex DC, Guise PG, Searle MP, Windley BF, Petterson MG, Jan MQ, Luff IW (1989) K-Ar and Ar-Ar geochronology of the Himalayan collision in NW Pakistan: constraints on the timing of suturing, deformation, metamorphism and uplift. Tectonics 8:881–909

Treloar P, Petterson MG, Jan MQ, Sullivan MA (1996) A re-evaluation of the stratigraphy and evolution of the Kohistan arc sequences, Pakistan Himalaya: implications for magmatic and tectonic arc-building processes. J Geol Soc Lond 153:681–693

Upadhyay R, Frisch W, Siebel W (2008) Tectonic implications of new U-Pb zircon ages of the Ladakh batholith, Indus suture zone, northwest Himalaya, India. Terra Nova 20:309–317. https://doi.org/10.1111/j.1365-3121.2008.00822

Van Haver T (1984) Etude Stratigraphique, Sedimentologique et Structurale d'un Bassin d'Avant Arc: Exemple du Bassin de l'Indus, Ladakh, Himalaya. Univ. de Grenoble, Grenoble, France, 204 pp

Wadia DN (1932) Notes on the geology of Nanga Parbat, Mt Diamir, and adjoining parts of Chilas, Gilgit district, Kashmir. Geol Surv India Rec 66:212–234

Walsh JMJ, Buckman S, Nutman AP, Zhou R (2019) Age and provenance of the Nindam Formation, Ladakh, NW Himalaya: evolution of the intraoceanic Dras Arc before collision with India. Tectonics 38:3070–3096. https://doi.org/10.1029/2019TC005494

Weinberg RF, Dunlap WJ (2000) Growth and Deformation of the Ladakh Batholith, Northwest Himalayas: implications for timing of continental collision and origin of calc-alkaline batholiths. J Geol 108:303–320

Wu FY et al (2007) Zircon Hf isotopic constraints on the sources of the Indus molasse, Ladakh Himalaya, India. Tectonics 26:TC2014. https://doi.org/10.1029/2006TC002051

Zeitler PK (1985) Cooling history of the NW Himalaya. Pakistan Tect 4:127–151

Zeitler PK, Chamberlain CP (1991) Petrogenetic and tectonic significance of young leucogranites from the NW Himalaya, Pakistan. Tectonics 10:729–741

Zhou R, Aitchison JC, Lokho K, Sobel ER, Feng Y, Zhao J-X (2020) Unroofing the Ladakh Batholith: constraints from autochthonous molasse of the Indus Basin, NW Himalaya. J Geol Soc 177:818–825. https://doi.org/10.1144/jgs2019-188|Vol

Chapter 3
Geology and Lithological Variations of the Ladakh Batholith Within the Leh–Ladakh Region

Abstract This chapter presents geological, lithological, and structural data from a 600 km^2 area of the Ladakh batholith, centred on Leh, extending from Taroo in the NW to Stakmo in the SE. The baholith comprises 80% plutonic rocks intruded into pre-batholithic, semi-pelitic to psammitic, turbidite schists and gneisses. Basic–intermediate gabbro–diorite and hornblendite rocks dominate the earliest stages of the batholith. The largest gabbro–diorite pluton extends over c. 100 km^2. Granodiorites and granites form c. 50–60% of the batholith. The largest plutons are 120–60 km^2 in area. Minor intrusions and mineral rich veins/lenses are prevalent throughout the area. Dykes and sills vary in composition from basalt to andesite to rhyolite. Tourmaline, epidote, chorite and sericite are common mineral vein material. The character of plutonic rocks varies between undeformed to cleaved, and highly-sheared, with gneissic foliation common in many areas. Metamorphism varies from very low grade to sericite-lower greenschist grade, to amphibolite and locally partial melt grade. The batholith and pre-batholithic meta-sediments are affected by numerous deformation events that have produced thrust-faulting, folding, shearing, and brittle high-angle normal and reverse faults. A sedimentation-magmatic-deformation-temporal model is presented which indicates: (1) early intrusion of gabbros and diorites with abundant semi-pelite xenoliths and mafic dykes; (2) granitoid plutons, sills, dykes and mineral veins intruding a gabbro–diorite-meta-sedimentary host; (3) shearing and brittle fault deformation with basalt, quart–porphyry and rhyolite dykes. A wide variety of photographic illustrations present the character, structure and lithological complexity, variation and heterogeneity from outcrop to whole-landscape scales.

Keywords Ladakh batholith · Plutonic rocks · Deformation

3.1 Batholith Overview in the Leh–Ladakh Region

This study focuses on the area shown in Fig. 3.1 and is referred to as the study area or Leh–Ladakh region. The area covered by Ladakh batholith rocks in the Leh–Ladakh region between Taroo in the NW and Stakna in the SE covers an area of c. 600

M. G. Petterson, *Himalayan Thick-Skin Basement Deformation of the Ladakh Batholith, Leh-Ladakh Region, NW India*, Springer Geology,
https://doi.org/10.1007/978-3-031-31566-4_3

km^2. The bulk of the batholith area (c. 80%) comprises a range of plutonic rocks which vary in composition between gabbro and granite (Fig. 3.1). Approximately 20% of the area comprises a range of meta-sedimentary rocks, largely semi-pelitic schists, which are the oldest rocks present in the batholith, being relics of the pre-intrusive sedimentary rock assemblage, now represented as contiguous stratigraphy, xenoliths within, and screens between, plutonic igneous units. The oldest igneous rocks are the gabbro-diorites which can be spatially-intimately associated with the meta-sediments: numerous outcrops are variations between dark, high-grade metamorphic schists and gneisses, and deformed gabbro–diorites. Some gabbro–diorites can contain high percentages of meta-sedimentary material, and relict bedded meta-sedimentary stratigraphy, up to hundreds of metres long and several to tens of metres wide. Higher silica plutonic rocks invariably intrude the more mafic plutonic rocks as sheets, major plutons, bosses, laccoliths, sills, and dykes. They vary in composition between granodiorites only slightly more evolved than quartz–diorites, to highly felsic granites with mafic minerals comprising only 5–10% of the rock. The average composition of felsic rocks straddles the granodiorite–granite boundary. Felsic plutonic rocks can also contain xenoliths of meta-sedimentary material, or gabbro–diorite, or largely assimilated and metasomatized (often hornblendized) material. Reaction rims, mm–cm thick, often occur at the edge of xenoliths, evidencing magma–xenolith geochemical interactions. The youngest plutonic rocks present within the batholith are minor intrusions and mineral veins that span the compositional range from basalt/micro-gabbro to quartz–feldspar porphyritic andesites/micro-diorites, and rhyolites/micrograntes. Minor intrusions are particularly common in the Taroo valley, although dykes and sills are present throughout the area. Many areas of the batholith are strongly deformed, with shear fabrics and/or fracture–cleavage structures being common. Plutonic rocks are locally gneissic where highly sheared, displaying well-developed mafic–felsic segregation foliations. Meta-sedimentary rocks vary in grade from sericite–greenschist to high-amphibolite, and occasionally migmatitic-partial melt grade. The south-eastern part of the area contains a high density of later to latest stage granite sills, laccoliths, stocks and bosses (Fig. 3.1). The general geological sequence observed within the batholith is: (1) deposition of pelagic to hemi-pelagic turbiditic muddy siltstones and fine-medium sandstones within an oceanic environment. Deposition was accommodated in part through half-graben style extensional faults; (2) intrusion of gabbros, diorites, and quartz–diorites as large plutons and small-scale veins, sills and dykes; (3) intrusion of predominantly coarse grained hornblende rich granodiorites, grading into a two-feldspar (white and pink) quartz-rich granite; (4) folding, shearing and thrust deformation phase; (5) intrusion of bimodal dykes and sills (basalt and rhyolite) syn-to early post-deformation; (6) intrusion of smaller scale granite plutons and low-mafic mineral micro-granite to rhyolite sills and dykes, together with quartz–porphyry intermediate and basalt dykes; (7) second stage of thrust to shear dominated deformation $\pm$ accompanying minor intrusions (mostly acidic, but some basic); (8) late stage reactivation of earlier structures with extensional and reverse fault motions and occasionally accompanied by minor intrusive activity. Before describing the details of the geology of different parts of the Leh Ladakh region, it is instructive

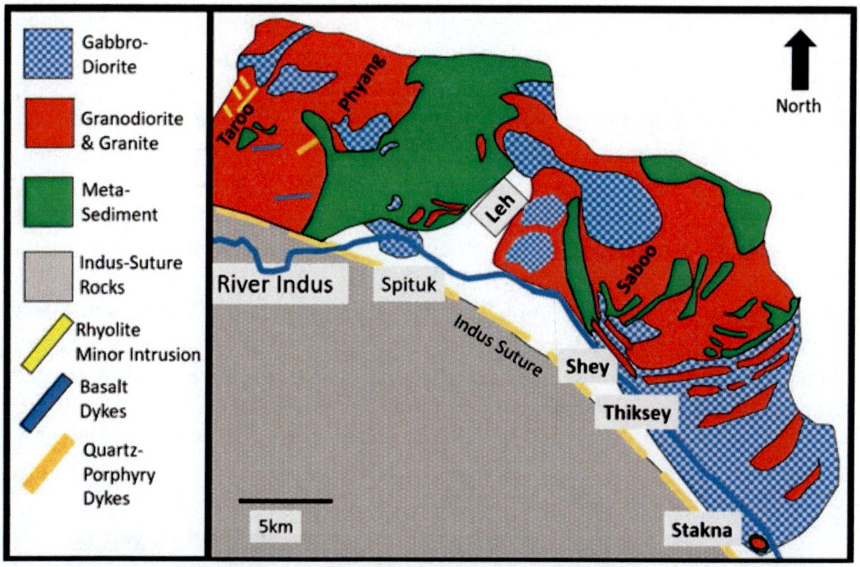

Fig. 3.1 Simplified geological map of the Leh–Ladakh region

to fully define the range of lithologies present within the batholith. Whilst setting out the various lithological components as discrete elements it is important to bear in mind that some outcrops are heterogeneous and can contain much of the variety described below in one outcrop

3.2 Meta-sedimentary Rocks

Meta-sedimentary rocks are the most enigmatic to fully describe, as outcrops can be patchy, remote, and difficult to access, or intruded by such a high density of plutonic rocks that the original character is lost. Meta-sedimentary sequences occur as relatively contiguous sequences of stratigraphy such as can be observed in outcrops between eastern Phyang and the Leh valley, or Upper Saboo (Fig. 3.1), or, more often, as significant but only locally (10–100s of m) contiguous within a meta-sedimentary-gabbro–diorite complex, or as xenoliths within, and screens between igneous intrusions. All meta-sedimentary rocks have undergone some form of metamorphism and deformation. The least metamorphosed and deformed outcrops, located within lower-strain zones (e.g. parts of the Sankar valley, Figs. 3.2, 3.3 and 3.4) retain a strong pelagic-hemi-pelagic turbiditic character, although metamorphic fabrics and minerals will still be present. The most highly metamorphosed and deformed meta-sediments are highly sheared, intruded, and recrystallised, exceptionally becoming mylonites or migmatitic partial melts (Fig. 3.5). The bulk of meta-sediments are metamorphosed to sericite–greenschist-medium-upper amphibolite grade and retain sufficient relict sedimentary characteristics to allow identification as meta-sediment. Locally, it can be difficult to distinguish meta-sediments from intruding gabbro–diorites.

Meta-sediments are mostly light–dark charcoal grey in colour, and pelitic-semi-pelitic to psammite in character. They retain aspects of an original sedimentary granular clastic texture with a high abundance of sub-rounded to sub-angular quartz grains. The most common minerals are quartz, hornblende, biotite, and plagioclase, with grain sizes varying between 1–2, and 6 mm. The more recrystallised meta-sediments lose aspects of any original clastic texture, which is replaced by crystalline feldspar, recrystallised metamorphic quartz aggregates, and hornblende–biotite crystals. At the highest strain levels secondary cleavage, schistose, and gneissic fabrics dominate, exceptionally presented as high-strain, small-grain size mylonites, cataclasite breccias, and migmatites, with in-situ partial melt felsic veins. Meta-sedimentary sequences are often caught up in thrust and shear zones, perhaps acting as a lubricating material.

Fig. 3.2 Relatively undeformed coarse greywacke sandstones (**a** and **c**) and fine-medium grained sandstones with mud flakes (**b** and **d**). Sankar. Pencil tip for scale (2 cm)

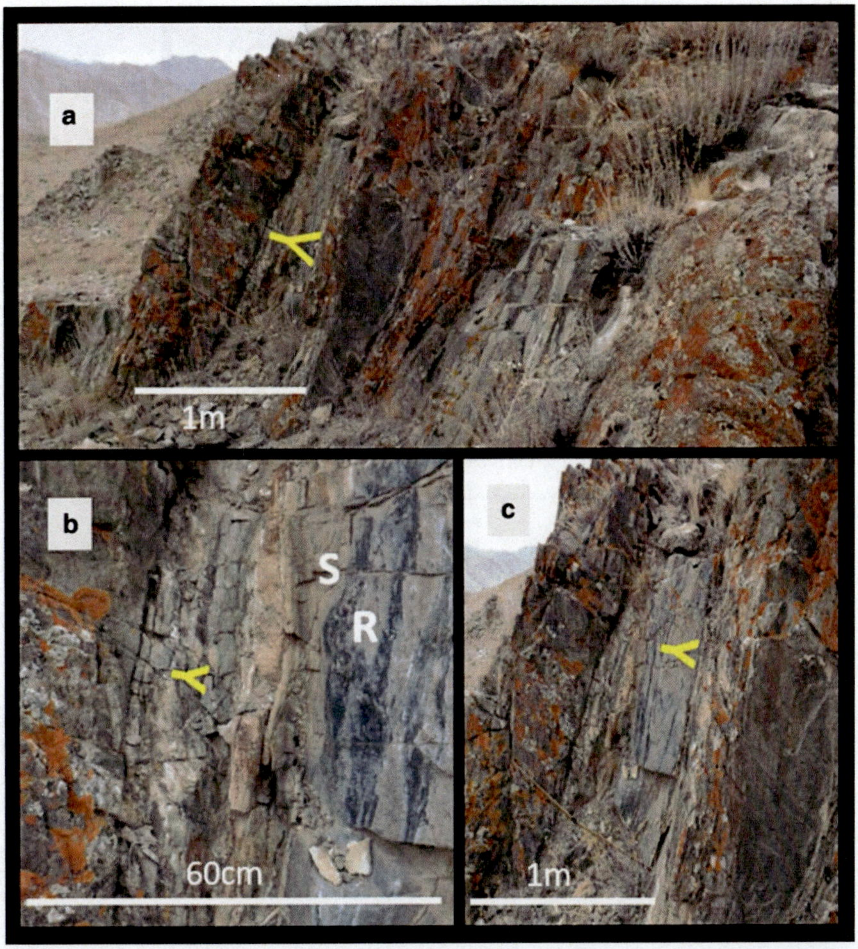

Fig. 3.3 Exposed sequences of relatively undeformed sub-vertically bedded turbiditic sandstones, siltstones, and mudstones. The sequence comprises c. 60% sandstones and 40% mudstones and siltstones. Bedding structures vary in thickness between 20 cm and over 1 m. Most beds are graded with coarse to medium sandstone bases and mudstone tops. Some units display erosional base scours (S) and upper ripple structures (R). Y indicates right way-up

A particularly common facies of meta-sediment occurs as xenoliths within gabbro–diorite and felsic plutons. Xenoliths are mostly elongated within shear zones and penetrative meta-plutonic fabrics, although occasionally they are more equant and sub-rounded in shape (Fig. 3.6). They mostly vary in length between a centimetre or less, ranging up to 1–5 m long in outcrop. At a larger scale they can extend tens to even hundreds of metres within plutons. Long: short axis ratios vary between 2:1 and > 20:1 though mostly they are c. 4:1–8:1. Some parts of the batholith have a remarkable stripey appearance with a high abundance of xenoliths within a host

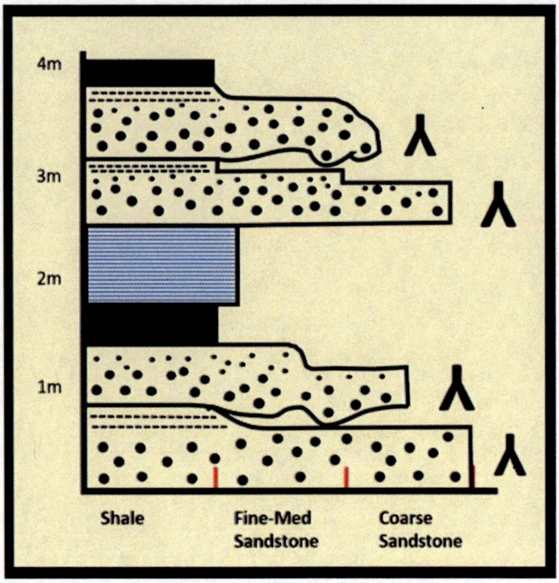

Fig. 3.4 Graphic log of c. 4 m thick turbidite sequence. Note graded bedding, basal erosional scours, and way-up criteria

pluton forming the most obvious component of gneissic fabrics, contrasting in colour with the paler plutonic rock. Xenoliths vary in appearance from obvious semi-pelite with a clastic character, to hornblende-rich recrystallised rocks, to barely visible ghost relics of former xenoliths. Xenoliths often have reaction rims, mm–cm thick around them, composed mostly of hornblende ± biotite, or silica and sometimes chlorite–epidote.

Figures 3.2, 3.3, 3.4, 3.5, 3.6 and 3.7 illustrate a range of lithological characteristics of the meta-sediments within the Leh–Ladakh region. Figure 3.2 presents examples of low-grade metamorphosed meta-sediments with only limited deformation. At one end of the spectrum these samples are coarse grained muddy sandstone (greywackes) with evidence of lamination-bedding, and poor sorting. At the other end of the spectrum the rocks are laminated fine-medium grained sandstones to siltstones with thin mud flakes. Figures 3.3 and 3.4 present photographs and a graphic sedimentary log from a c. 4 m sequence of vertically bedded turbiditic sedimentary rocks. The rocks vary in grain size from coarse grained sandstone to shale. Many units display way up criteria in the form of normal graded bedding, erosional scours, and sedimentary ripple structures. Some units have a medium-coarse to very coarse-grained sandstone base that grades upwards through fine grained sandstone and siltstone, with mudstone-shale parallel-laminated units at the uppermost parts of the bed. Some beds are almost entirely siltstone to shale/mudstone. Bedding thicknesses vary between c. 10–20 cm and 50–70 cm. Sedimentation is rhythmic. These units are interpreted as turbidite sequences, with the lack of very coarse materials (conglomerates, pebbly

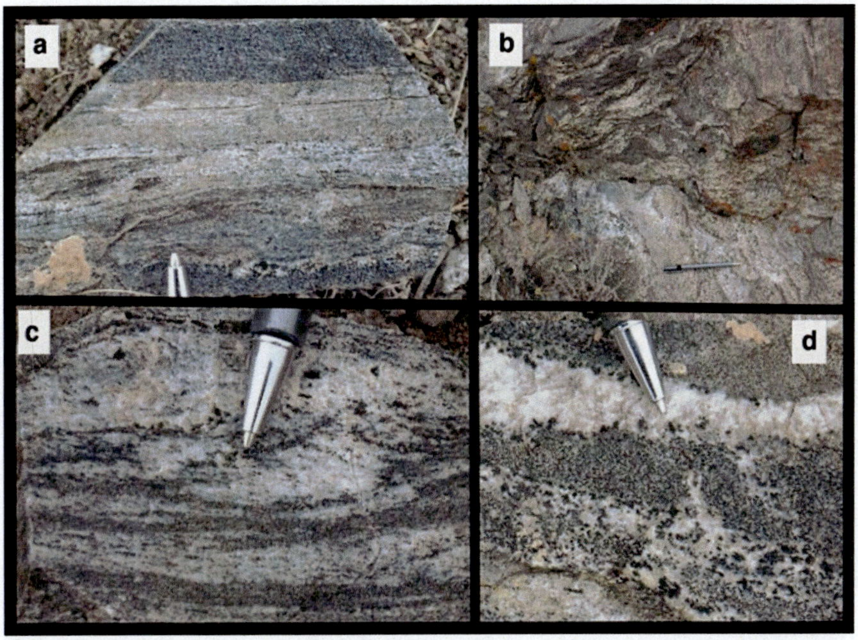

Fig. 3.5 High grade amphibolite to migmatite grade meta-sediment, Sankar. *Note* Bedded nature of **a** with relict granular clastic texture alongside recrystallised and lit par lit micro-granite sheet; migmatitic facies in **b** with numerous microgranite lenses; and coarse grained recrystallised meta-sediment with partial melt quartz–feldspar layers in **c** and **d**. Wave-like structures in **c** close to pencil tip may be relict cross-lamination structures. Pencil tip for scale (2 cm)

sandstones) and the abundance of finer grained sequences (siltstones and mudstones), together with the thin-moderate bedding thicknesses, suggesting a distal to medial pelagic to hemi-pelagic turbiditic setting, deposited within a deep marine basin.

Figures 3.5 and 3.6 present higher grade meta-sedimentary units, up to migmatite grade. The outcrops display strong gneissic to schistose foliations with medium-coarse to very coarse-grained recrystallised textures. Melanosome–leucosome segregation layers are well-developed. Larger hornblende porphyroblasts stand proud of the metamorphic textures, some of which have rotational syn-tectonic snowball textures, and others appear post-tectonic, cutting the bulk of penetrative fabrics. Some migmatitic facies comprise of over 50% felsic–quartz silicic in-situ partial melt lenses, veins and sheets. One facies retains curvilinear structures, which could be purely metamorphic, or could also reflect relict cross-bed ripple structures.

Figure 3.6 presents a range of xenolith morphologies within unstrained and strained rocks. Unstrained xenoliths are more equant with low length/width ratios and can be only moderately altered retaining much their original lithological character. Other xenoliths are extended with length/width ratios up to 20:1 or more and are highly altered. Some lithologies are relatively free of xenoliths, although this is rare: most plutons contain c. 5–15% visible xenoliths, whilst exceptionally other

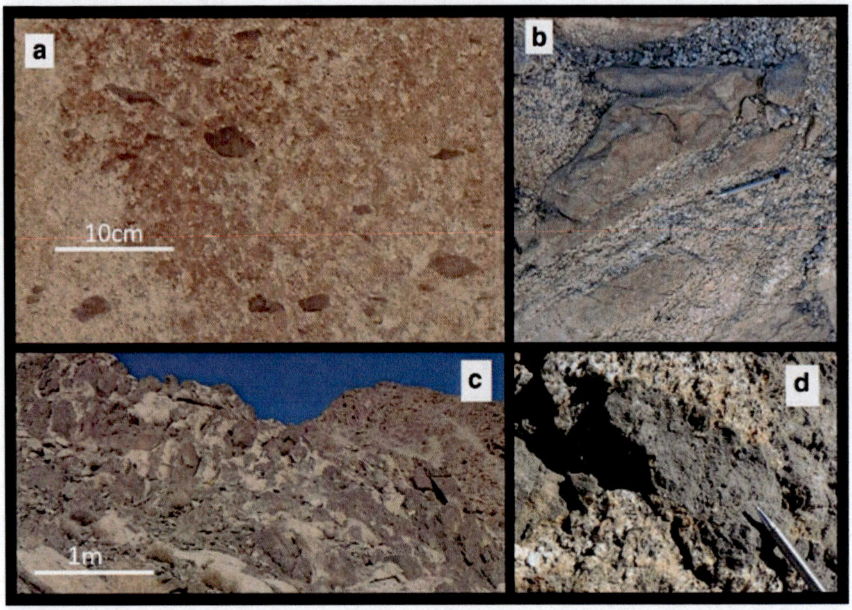

Fig. 3.6 Meta-sedimentary xenoliths (Taroo). Xenoliths in **a** and **d** are equant, relatively undeformed, and form c. 10–15% of the rock outcrop. **b** is a close up of facies **c** which contains > 70% xenoliths, some of which are meta-sedimentary and are aligned parallel to shear fabrics. Pencil for scale in **b** and **d** (15 cm)

Fig. 3.7 Tectonised Meta-sediments. **a** Presents a tectonised-cleaved-sheared meta-sediment unit below the syn-deformational microgranite fish/lozenge. Facies varies between rock flour-cataclasite-fine fault breccia and mylonite. Thiksey. Notebook for scale (17 cm) **b** illustrates a tectonised–mylonitised meta-sediment with strong planar and linear fabrics, and grain size reduction, together with silicification (Stakna). In both cases the meta-sediment is spatially associated with, and facilitates shearing and thrust-fault movement

units may contain up to 70% xenoliths. Meta-sediments are commonly associated with thrust faults and shear zones and may play a lubricating role, particularly in some of the higher-strain structures. Mylonites are silicic, strongly foliated, and lineated rocks with a fine-grained texture and highly sheared fabrics. Cataclasites, and rock-flour/breccia, display highly sheared, cleaved, and comminuted textures.

3.3 Gabbro–Diorites

Gabbro–Diorites occur as extant small-large plutons across the region, being particularly abundant in the SE of the region. They also occur as xenoliths within younger felsic intrusions. The largest pluton is the c. 100 km^2 Thiksey Gabbro–Diorite, followed by the c. 30 km^2 Leh–Saboo Diorite, and smaller ones around Spituk, and in the Phyang–Taroo region (Fig. 3.1). The most common minerals present are hornblende, biotite, plagioclase feldspar and quartz. Ratios of hornblende + biotite: feldspar: quartz vary between c. 60:35:5 and 30:55:15. Gabbro–diorites are mostly medium-coarse grained rocks, with average grain sizes of 4–7 mm. Hornblende is often the largest crystal extending up to exceptionally 2–3 cm, though normally between 8 and 12 mm in length. Hornblende is the most common phenocryst phase and is often euhedral–subhedral with square and rectangular cross sections and 120° cleavage intersections: rhombohedral cross sections are visible within phenocrysts. Biotite accompanies hornblende or is closely spatially associated. Sometimes hornblende and biotite crystals display an acicular–decussate habit, and within other outcrops hornblende can form glomerocrystic aggregates. Feldspar is the next largest crystal, displaying euhedral–subhedral, rectangular, and square cross-sectional crystals, that vary between 5 and 8 mm in length, exceptionally up to 10–12 mm. Occasionally feldspar occurs as elongate lenticular semi-crystalline forms within the more highly sheared plutons. Quartz forms the smallest grains, often occurring as glomerocrystic aggregates, comprising numerous smaller polygonized grains. Quartz aggregates vary in shape, can be aligned to any penetrative fabric, or form semi-amorphous clusters or blebs 3–6 mm in size. Exceptionally pegmatitic gabbro facies are present with crystals up to 3 cm.

Gabbro–diorites are highly variable in character throughout the region, from dark, melanocratic, grey-black hornblende rich gabbros, to lighter mesocratic quartz-rich Quartz Diorites. Individual outcrops can contain abundant felsic intrusions which break up the darker gabbros, and there can be compositional variations across outcrops from more mafic to more felsic. meta-sedimentary and gabbroic xenoliths are common in many outcrops. Later felsic intrusive rocks can break up the continuity of gabbro–diorite outcrops creating 'mixed pluton' assemblages.

Gabbro–diorites also vary in terms their presence/absence or levels of development of secondary fabrics. Many outcrops are deformed to some extent, and secondary fabrics are common. These vary between outcrops with well-developed contiguous cleavage planes and fracture, to ductile penetrative shear fabrics to even mm scales. Some gabbros are gneisses with melanosome–leucosome segregations and metamorphic recrystallisation and shearing fabrics dominating.

Figures 3.8, 3.9 and 3.10 illustrate a range of gabbro–diorite lithologies. Figure 3.8 illustrates the more melanocratic character to the gabbros, some facies being dark grey in colour (Fig. 3.8a, b), with up to 60% mafic minerals, predominantly hornblende. Grain size varies from medium-coarse grained. Hornblendes tend to be the larger crystals and can form porphyritic textures locally. Figure 3.8b displays a gabbro outcrop riddled with granitic veins and sheets that form a dense anastomosing network and break up the host gabbro. Figure 3.8c illustrates aspects of a medium grained gabbro with local pegmatite veins of hornblende–feldspar–quartz. Figure 3.9 presents a range of diorite lithologies, from more melano-diorites to leuco-diorites that border on granodiorite in composition. Figures 3.9b, 3.10b and c illustrate gabbro–diorite gneisses with well-developed melanosome–leucosome segregation foliations and stretched xenoliths. Deformed plutonic rocks are common in the Leh–Ladakh region. Figure 3.10a presents a mountainside image to give perspective on the scale of some gabbro intrusions. The gabbro is intruded by numerous post-gabbro granite sheets, bosses and laccoliths.

Fig. 3.8 Leh–Ladakh Region Gabbros. **a** From Saboo is coarse grained hornblende-rich gabbroic facies with larger hornblende phenocrysts. **b** From Phyang is a medium grained hornblende-rich melano–gabbro intruded by anastomosing granitic sheets and veins. **c** is from Taroo and is a part pegmatitic facies with lenses and 'pipes' of very coarse hornblende–biotite–feldspar–quartz crystalline assemblages

Fig. 3.9 Leh-Ladakh Region Diorites. **a** and **c** from Spituk are coarse grained diorites within close proximity of one another showing local variations between more and less mafic varieties of hornblende-plagioclase-quartz dominated lithologies. **b** is from Shanti Stupa and displays a more leuco-dioritic facies with a strongly foliated gneissic texture. Pencil (2 cm) for scale.

Fig. 3.10 **a** From Phyang presents a mountainside view, 250 m of vertical topography, with gabbro–diorite dominating the terrain, intruded by granitic sheets. **b** and **c** are from the Shanti Stupa area and present a strongly gneissic gabbro, (**b**) and a Quartz–Diorite (**c**) respectively. Note the melanosome–leucosome segregated foliations, and stretched xenoliths, all aligned to the gneissic foliation

3.4 Granodiorites–Granites (Granitoids)

Granodiorites–granites (granitoids) occur as larger plutons as well as laccoliths, stocks and bosses, and sills and dykes of various scales. Two large granitoid plutons dominate the study area: the c. 120 km^2 Saboo Granitoid Pluton and the c. 60 km^2 Taroo–Phyang Granitoid Pluton (Fig. 3.1). The most common minerals present are white (plagioclase) and pink–red (alkali) feldspar, quartz, biotite, and hornblende. Hornblende + biotite: feldspar: quartz ratios vary between 20–25%: 20–25%: 50–60% in the more mafic granodiorites and 10%: 30–40%: 50–60% in the more felsic granites. Pink feldspars are locally common, often display Carlsbad twinning, and can comprise a significant proportion of the feldspar population. Muscovite is present in a minority of granitoid outcrops. Granitoids are mostly medium-coarse grained, with average grain sizes of 5–15 mm. Hornblende forms the larger crystals and is mostly euhedral, displaying square–rectangular sections, up to 20 mm long, but mostly 5–12 mm in length. Feldspars are euhedral–subhedral crystals, mostly displaying rectangular or square shapes, measuring up to 15 mm, but mostly 5–8 mm in length. Quartz is the last crystallising stage, tending to occupy whatever space remains within the rock matrix. Quartz can form blebs or elongate–oval shapes, and is often glomerocrystic, with larger quartz aggregates comprised of smaller, polygonised sub-grains.

Granitoid rock fabrics vary between crystalline, granular, even sugary and undeformed, to highly sheared and cleaved gneisses, with leucosome–melanosome segregation foliations. Meta-sedimentary, and gabbro–diorite xenoliths, are common, as are felsic, intermediate and basic minor intrusions. Some facies of granite weather a strong red to rust-brown colour and attract lichen growth that has a similar colouration.

Figures 3.11 and 3.12 present a range of lithological characteristics of the Leh–Ladakh granitoids. A range of facies from coarse grained granular granites with euhedral interlocking grains, and the presence of pink and white feldspar, alongside milky-white quartz and a smaller proportion of mafic minerals is presented in Figs. 3.11b, c and 3.12b. Paler coloured medium-coarse grained granodiorites with a higher proportion of hornblende and biotite minerals, some of which are large and porphyritic, with interlocking granular white feldspar and quartz are presented in Fig. 3.11c and f. Finer grained granites are presented in Fig. 3.11a and e. A highly cleaved granite, most typical of the high strain areas of the North–North-eastern Zones is presented in Fig. 3.12b, and a strained, gneissic granodiorite with elongate abundant xenoliths is presented in Fig. 3.12c. Granite rock faces are often used for carvings of local Tibetan Buddhist deities in this region: Fig. 3.12a presents such as case for the deity Avalokteshivara (Chambak) in south Taroo.

3.5 Minor Intrusions and Assorted Minerals Veins

The Ladakh batholith in the Leh region hosts several minor intrusive phases and mineral veins. Minor intrusions span the compositional field from basalt to andesite through to rhyolite. Hydrothermal and metamorphic minerals can occur within thrusted, cleaved, and sheared rock, including clay minerals, tourmaline, feldspar, epidote, chlorite, and quartz. All areas of the batholith host at least a small percentage of minor intrusions. Taroo has several areas with relatively abundant dykes and sills, spanning the full compositional range, and areas such as Agling to Saboo and Stakmo, host a relatively large number of granitic and microgranitic veins, dykes, sills, laccoliths and stocks.

Minor Intrusions are pre-, syn- and post the main deformation of the Ladakh batholith. The oldest minor intrusions are mostly highly disrupted and often unrecognisable as original dykes or sills. This phase is predominantly basaltic, with dykes and sills 10 cm–1 m wide. Dykes are mostly sheared and disrupted/broken up and may be represented almost entirely as basaltic xenoliths within sheared plutonic rock or be broken by magmatic flow within plutons. Syn-deformational minor intrusions are mostly microgranitic–rhyolitic, although some basic dykes exist. They intrude as sills, dykes, veins, lenses, and pods. Most dykes and sills are cm–1 m wide although some are several metres to 10s of metres wide. Rhyolitic, elongated, lensoid tectonic

Fig. 3.11 Granodiorites–Granites. **a** From Shey displays a feldspar–quartz leucocratic medium grained and tectonised granite. **b** From Phyang displays a coarse-grained granular facies granite with pink and white feldspar, and large quartz crystals. **c** From Taroo is a granodiorite with a higher proportion of hornblende–biotite mafic crystals, some of which are elongate and porphyritic and some of which have embayed crystal margins. The leucocratic component comprises coarse crystalline feldspar (some twinned) and quartz. **d** From Phyang displays the contact between a coarse granite with a hornblende–biotite-pink and white feldspar–quartz mineralogy and a semipelite; **e** from Sankar displays a medium grained granite with c. 15% mafics and **f** from Taroo presents a medium-coarse grained granular hornblende–granodiorite. Pencil and notebook for scale (pencil is 15 cm long, tip 2 cm, and c. 5 cm of notebook exposed in **d**

Fig. 3.12 **a** From Taroo displays a carved Avalokteshivara (Chambak) deity within a coarse-grained, cleaved granite rock face. **b** From Leh valley presents a strongly cleaved granite and **c** from Agling, south Leh, presents a strongly foliated granodiorite with abundant stretched xenoliths

'fish', are present within some shear zones. Millimetre sized veins are also locally present. Syn-deformational intrusions are often cleaved, particularly at the minor intrusion-host-rock interface, and they can form pinch and swell-boudinage structures. Post-deformation minor intrusions are uncleaved to moderately cleaved, have a fresh appearance, and extend contiguously across an outcrop and/or mountainside. Examples of basalt, andesite, and rhyolite are present: many are affected by latest stage, brittle, mostly high-angle, faulting.

Mineral veins are mostly mm–cm wide, exceptionally up to 1 m wide, and may also exhibit a disseminated habit as anastomosing small veins, individually mm–cm wide throughout the host rock close to shear zones and cleavage zones. Veins can be mono-mineralic, or have a vein symmetry with one vein at the centre grading into another mineral at the edge of the vein. Minerals such as epidote, feldspar, and silica–quartz, can form well-developed slickenside structures.

Figures 3.13, 3.14, 3.15, 3.16, 3.17 and 3.18 illustrate some of the variety of minor intrusion and vein features and lithologies. Figure 3.13 provides some spatial context for later stage basic and intermediate dykes, which can intrude across significant distances. These intrusions tend to be less altered and metamorphosed, and display original unaffected igneous textures and minerals, although intrusion margins are affected by cleavage. This figure illustrates how late-stage faulting impacts these youngest intrusions. Figure 3.14 gives a close-up of a basalt dyke and its chilled margin with the adjacent granite pluton. This dyke is c. 50% hornblende and feldspar, with ophitic and acicular mineral textures. Figure 3.15 illustrates aspects of a local rhyolite–microgranite intrusive centre in west-central Taroo with a local high abundance of micro-granite dykes, sills and veins measuring from a few cms to a metre wide up to c. 10–20 m. The intrusions can form prominent bright-white escarpment ridges and related topographic features that strike up hillsides. Although they are late-stage intrusions, parts of the rock mass are cleaved. Figure 3.15 also presents a coarse pegmatitic acid vein closely associated with shearing, which although cutting gneissic penetrative fabrics in the host granite, associated with the main most intensive deformation event, they also intrude into cleavage-zones formed by later thrust faulting deformation. Figure 3.16 presents an early basic dyke. Parts of the original dyke with sharp margins are preserved, although the dyke has been cut by shearing and faulting (or possibly through magmatic-related disruptions, Kumar 2020). Adjacent angular and abundant basaltic xenoliths evidence the fate of many of the early basalt dykes which are torn apart by thrusting, shearing, intrusive dynamics or magmatic processes. Figure 3.17 presents epidote–chlorite, tourmaline–quartz greisen assemblages and feldspar-white clay mineral veins. These are linked to deformation in the form of cleavage and shearing/thrusting, and associated hydrothermal activity, depositing secondary minerals within the sheared rock masses. One epidote outcrop displays well-developed slickenside structures indicating that these minerals were deposited during syn-deformation times. Figure 3.18 presents lithological examples of the later stage fine grained micro-granites and intermediate quartz–porphyry

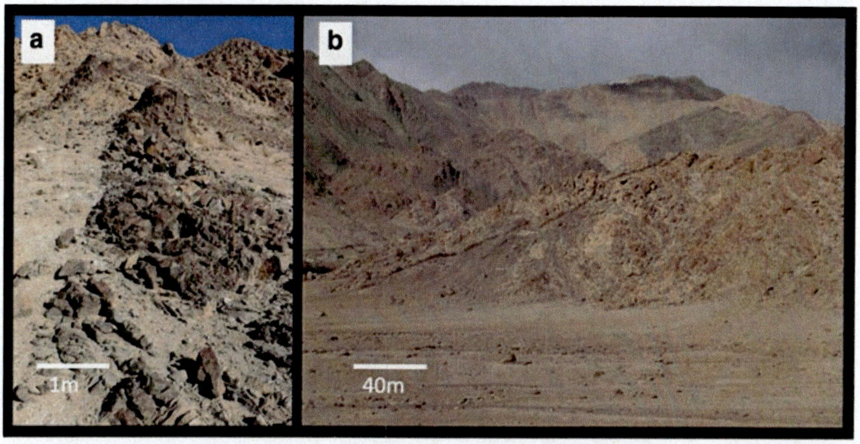

Fig. 3.13 Quartz–Porphyry andesitic dyke intrusions, Taroo. **b** Provides the overall context with the Quartz–Porphry dyke, cutting a predominantly granitic–granodioritic host rock over an extensive area. **a** Shows local detail with the darker quartz–porphry dyke striking up the mountainside, affected by late-stage brittle faulting. Although the dyke is mostly post-main deformation, it is cleaved, suggesting it intruded towards the end of the main deformational period

dykes. Quartz porphyry dykes tend to have a fine-grained matrix which can weather a sandy-brown colour but is mesocratic medium-grey when fresh. Key phenocryst phases are quartz, in the form of clear-milky aggregated 'bleb' lozenge structures, smaller equant-rectangular to bladed hornblende–biotite crystals, and larger feldspars that are either rectangular–square or display rounded, orbicular, zoned structures.

3.6 Larger Scale Examples from the Batholith

Previous sections have divided the batholith up into its various constituents based on lithological composition. This section presents examples of a broader scale appearance of the batholith. Arid high mountain environments like Ladakh present excellent opportunities for relatively uninterrupted views of batholiths. The full range of batholithic character from the first deposition of meta-sediments, through the main gabbro to granite plutons, and later minor intrusions along with the full range of deformation in the form of folding, cleavage, thrusting, and shearing are exposed. Figures below illustrate aspects of this geological diversity.

Fig. 3.14 Late-stage basalt (hornblende–feldspar) dyke cuts local granite intrusion, Taroo (left-hand-side). The dyke is up to 1 m wide in parts and displays a chilled margin. The rock displays ophitic textures with local acicular hornblende and feldspar crystals (right-hand-side). Pencil (15 cm) and pencil tip (2 cm) for scale

Figures 3.19, 3.20 and 3.21 are from Taroo valley. Figure 3.19 presents a photomontage looking northwards from the south-central part of the valley. The vertical relief present in the image is c. 500 m. This scene is reasonably typical of the N–NE Zone of the batholith within the NW part of the Leh–Ladakh area. The foreground granite is cleaved and has been affected by a degree of thrust-faulting. Cleavage planes control local topography producing a range of scales of benches, steep cliffs, and escarpments. The granite is cut by the youngest lithological unit in the area: a quartz–porphyry dyke which is itself faulted. Much of the mid-higher ground is composed of gabbros, granites, and semi-pelitic schists, the latter of which are strongly foliated and form a strong topographic bench topography. Part of the highest ground is composed of granite. The schist–gabbro–diorite–granite-late dyke temporal sequence is typical of the batholith. Figures 3.20 and 3.21 present detailed field relations of the youngest quartz–porphyry and basaltic dykes respectively which cut most structures but are also themselves cleaved and affected by late high angle brittle faulting. They probably intruded syn-post later stage thrust deformation.

Fig. 3.15 Western-Central Taroo hosts a relatively high density of thick (4–15 m) leucocratic dykes and sills that intrude over an extensive area (**a**, **c**). Whilst cutting most structures and lithologies they are locally cleaved, particularly close to intrusive margins. Some sills form escarpment features as in (**a**). **b** is from Phyang and displays a pegmatitic quartz–feldspar–biotite acid vein which cuts the local gneissic foliation intruding along cleavage planes that are part of later thrust deformation events. Pencil (15 cm) for scale in (**b**)

Figures 3.22, 3.23, 3.24, 3.25 and 3.26 present a series of interpreted photomontages from the NW to SE, covering the areas around Spituk and Leh (Fig. 3.22), the Leh valley area (Figs. 3.23 and 3.24), Stakmo valley (Fig. 3.25) and a view looking NW from Stakna (Fig. 3.26). Figures 3.22, 3.25 and 3.26 illustrate the area of the batholith strongly affected by a high density of later-latest granitic intrusions. The intrusions occur largely as sills of varying thickness and length, up to tens of metres thick and stretching several kms, and associated small plutons, dykes, laccoliths, and stocks. Most of these sheets are pre-latest deformation as they are locally folded, sheared, and thrusted, but are unaffected by earlier deformation events. Some sheets are syn-thrust to post-thrust in age. Figures 3.23 and 3.24 are from the Leh valley area and present larger plutons, again with the younger granites cutting the older gabbro–diorite-semi-pelite schists. Some plutons are not so easy to categorise as they can span a wide range in composition. The granodiorite with diorite pluton shown in Fig. 3.24 is an example of such an intrusion.

Fig. 3.16 Leh valley. Central dark fine grained basalt dyke cuts host granodiorite. Note how the dyke is faulted, cleaved, veined, and fractured. Abundant adjacent angular basaltic xenoliths within the host granodiorite evidence the disruption and break up of early dykes such as this one, potentially in-part through magmatic flow processes (e.g. Kumar 2020). Ruler (25 cm) for scale

3.7 Summary of the Key Aspects of the Leh–Ladakh Batholith and Intrusive-Deformation Model

Evidence provided in this section clearly presents: (1) the overall character of the batholith and its pre-intrusive basement geology; (2) aspects of the deformation of the batholith, particularly with respect to the production of well-developed schistose and gneissose foliation fabrics, and (3) the chronological sequence of intrusive activity within the batholith. The two following sections examine details of the deformation of the batholith. This section draws aspects of these strands of evidence and presents a synthetic overview. Sections below will demonstrate that the batholith has been deformed through (1) shallower ductile–brittle transition zone thrust faulting; (2) deeper, mostly ductile-level, but transitioning into the lower brittle zone, folding and shearing deformation, and (3) latest stage higher-angle faulting which is mostly extensional, but also compressional locally, and re-activation of older thrust faults.

Fig. 3.17 Mineral veins. **a** (Leh valley) and **b** (Spituk) present epidote–chlorite mineralised cleavage planes, with (**a**) displaying well-formed slickenside structures. **c** and **d** (Shanti Stupa area) display shear zones intruded by a greisen tourmaline–quartz assemblage in (**c**) and clay–alkali feldspar assemblage in (**d**). Pencil (15 cm) and compass-clinometer (10 cm) for scales

Figure 3.1 presents a synthetic overview of various geological-deformation aspects of the Ladakh Batholith in the Leh area. Figure 3.1 shows how much of the batholith comprises gabbro–diorite and granodiorite–plutons which intrude a schistose oldest meta-sedimentary unit. Figure 3.1 together with Fig. 3.27 also present those parts of the batholith that contain areas of relative high-density later-granitic sill intrusion. In Taroo this is constrained largely to the western-central parts of the valley with metres-tens of metres thick micro-granitic, rhyolitic, and fine grained, sills, dykes, and sheets cutting all other batholithic units (e.g. Figs. 3.15 and 3.16). High densities of granitic sills which are themselves locally folded, thrusted and sheared, but also locally cut the earlier main deformation structures are present mainly in the SE of the area, e.g. around Stakmo (Figs. 3.25, 3.26 and 3.27). Structures vary from being highly thrust dominated and highly cleaved at the southwestern margins of the batholith, and fold-shear dominated further north–northeast (see Chaps. 4 and 5 below for details). Figure 3.28 summarises the main aspects of the field relations of the batholith. Earliest medium-large scale gabbro-diorite plutons and accompanying basic dykes intrude a pre-existing turbiditic sandstone–siltstone basement. Plutons contain turbiditic meta-sediments as xenoliths (Fig. 3.28a). Later granitic plutons cut the earlier gabbro–diorite–basalt dyke sequence and themselves assimilate schist, gabbro–diorite and basalt xenoliths (Fig. 3.28b). Latest stage significant volumes of granitic sheets intrude the SE part of the region, and the north-Spituk-Taroo areas

Fig. 3.18 Upper fine-grained microgranite and lower quartz–porphyry (Q–P) intermediate dyke, Taroo. Q–P contain large quartz blebs, orbicular, rounded, zoned feldspars, and small hornblende–biotite crystals set in a fine-grained matrix

(Fig. 3.28c). A range of dykes from basalt–andesite–rhyolite intrude (Fig. 3.28c). Some of this later suite intrude during latest deformational to post deformational times but are themselves cut by the final high angle brittle fault stage.

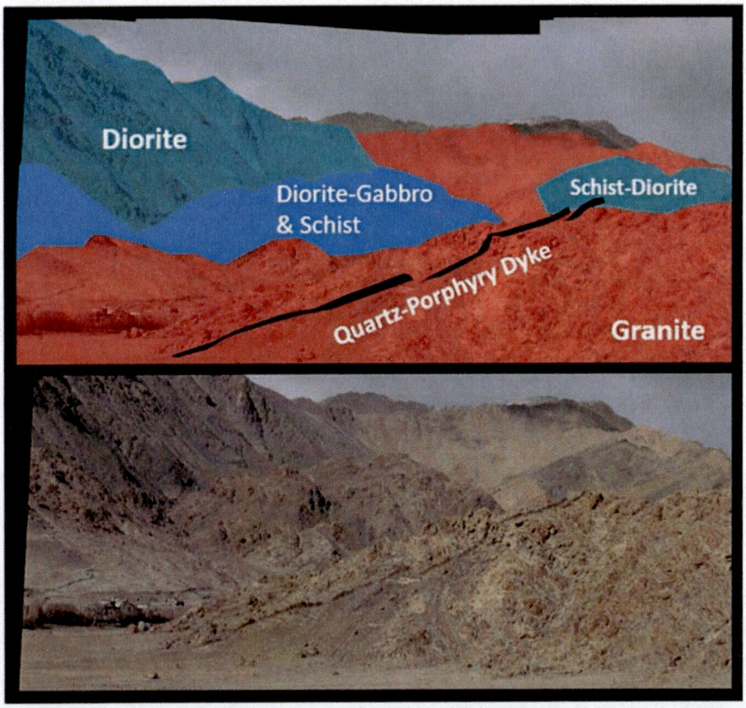

Fig. 3.19 Taroo valley. Older diorite and gabbro–diorite plus schist xenolith plutons are cut by younger granites and a youngest quartz–porphyry dyke. The granite forms the lower hills and is moderately cleaved forming tor-like and benched topography. The late dyke forms a prominent strike-bench feature. The middle ground is composed of bedded schists with dioritic and gabbroic intrusions: note how the schistose foliation strongly influences the bench-like geomorphology in the centre of the image. The highest mountains are c. 4400 m above sea level, 4–500 m above the valley floor, and are composed of diorite cut by granite. Houses for scale in the valley floor

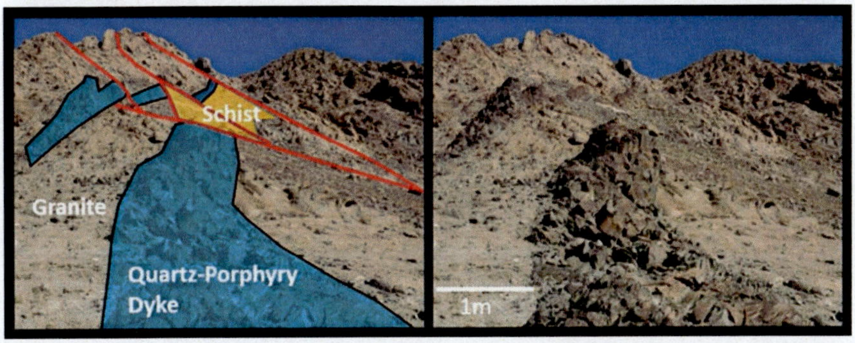

Fig. 3.20 Detail of the dyke presented in Fig. 3.21. Taroo valley. The Quartz Porphyry dyke, c. 1–3 m wide cross-cuts S_1 cleavage fabrics in the host granite but is itself cut by late brittle faults. The dyke is strongly jointed reflecting the presence of S_2 cleavage fabrics. Note the well-formed bench topography in the granite controlled by S_1 cleavage planes

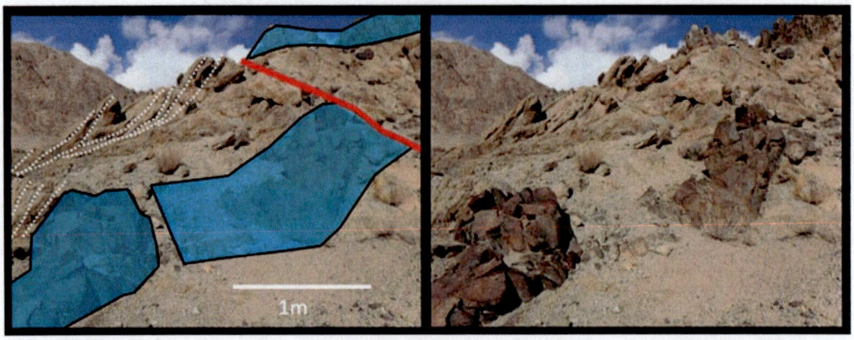

Fig. 3.21 Taroo. A late basic dyke, syn-post thrusting (dashed white lines show cleavage planes formed by thrust deformation) is cut by later brittle faulting

Fig. 3.22 View looking north-northeast towards Leh Town, Leh Airport in the foreground for scale. An older sequence of gabbro–diorite plutons, themselves intruding a meta-sedimentary basement, are cut by later granitic sills, stocks and laccoliths. Highest mountains are c. 5600 m above sea level, 2200 m above the airport valley

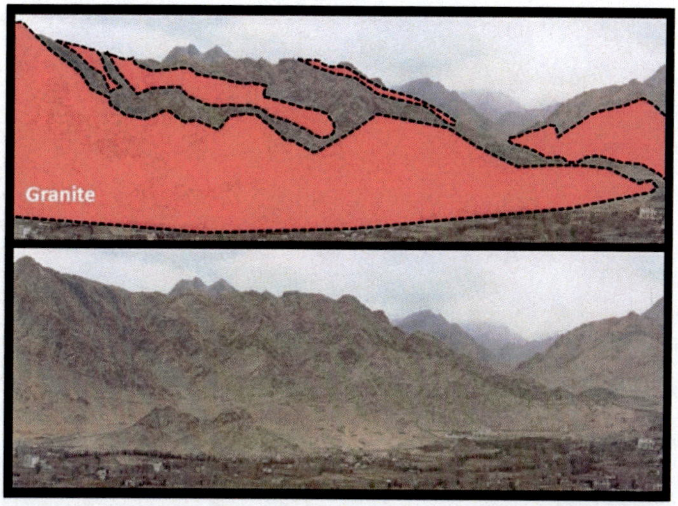

Fig. 3.23 North Leh valley and Sankar. Two larger granite plutons and smaller granitic sheets cut oldest dark gabbro diorites and meta-sedimentary schists. Villages, fields, and houses for scale in the foreground. Mountains are c. 4000 m above sea level, some 400–500 m above the valley floor

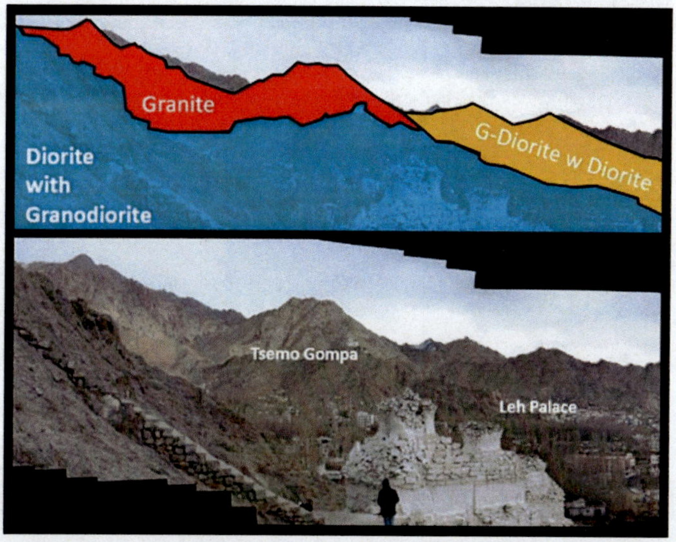

Fig. 3.24 View looking east from Shanti Stupa area towards the Tsemo Gompa and Leh Palace. The area around Shanti Stupa is underlain by highly sheared gabbro–diorite and meta-sediment: this geology extends to the Tsemo Gompa and Leh palace areas which stand on top of thrust-ridges that themselves rise above the Leh valley. Lighter, more granodioritic rocks occupy the central-northern mountains in the mid-left of the photo-montage. Highest mountains are up to 4500 m high, some 1000 m above the valley floor

Fig. 3.25 Prolific sheets, sills, and laccoliths intrude an older gabbro–diorite-meta-sedimentary schist basement, Stakmo valley. The highest mountains in the distance are 5–7 km away and rise to 4600–5000 m above sea level, or 1000–1500 m above the valley floor

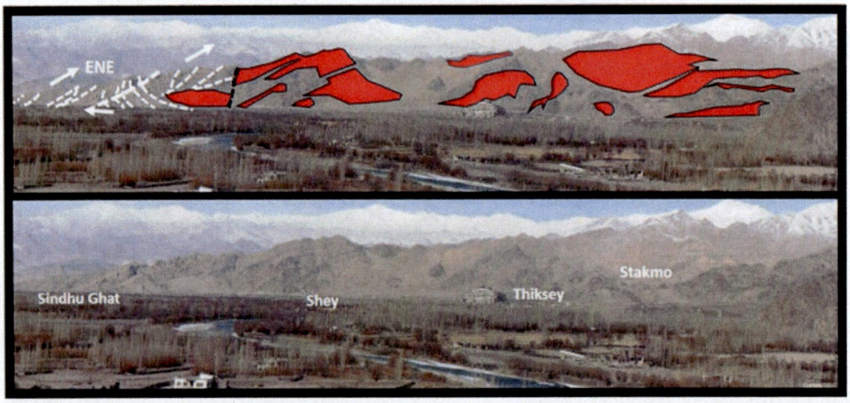

Fig. 3.26 View looking NW from Stakna Gompa, River Indus and its valley in the foreground. Later granitic plutons, stocks and sills intrude older darker gabbro–diorite and schist basement around Stakmo and north of Shey and Thiksey. Thrust-backthrust and recumbent fold-nappe structures are present on the left-hand side of the photomontage in the Sindhu Ghat area. The River Indus is at a height of c. 3200–3300 m above sea level and the distant snow-capped mountains are up to 5600 m above sea level

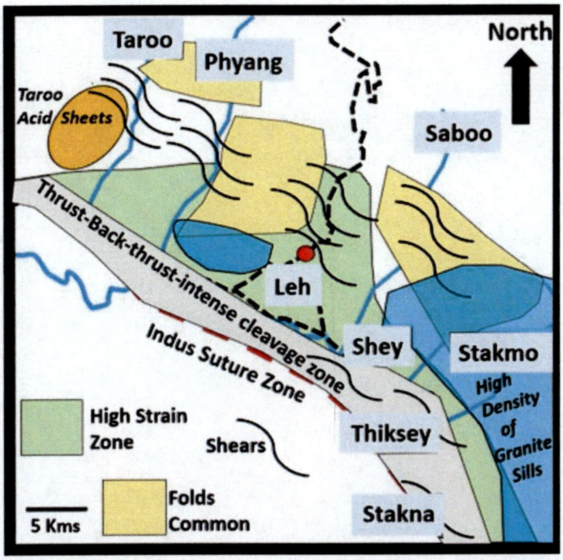

Fig. 3.27 Synthetic overview geo-structural map of the Leh–Ladakh area batholith. Details of the structural domains are discussed in the two structural sections below. The green zone is the High Strain Zone (HSZ) with a more heterogeneously strained zone NE of the HSZ (see Chaps. 4 and 5 below for details). The HSZ is particularly highly strained and thrust-faulted at the southwestern margin. Folding becomes the more dominant structural style in the north of the region. Shear deformation is prominent within a NW–SE band some 10–15 km NE of the southernmost margin of the batholith, and locally within the HSZ. Western Taroo is the centre of a high density syn-post deformation acid sheet zone. High densities of granitic sill intrusions are present in the central-south and south-eastern parts of the area

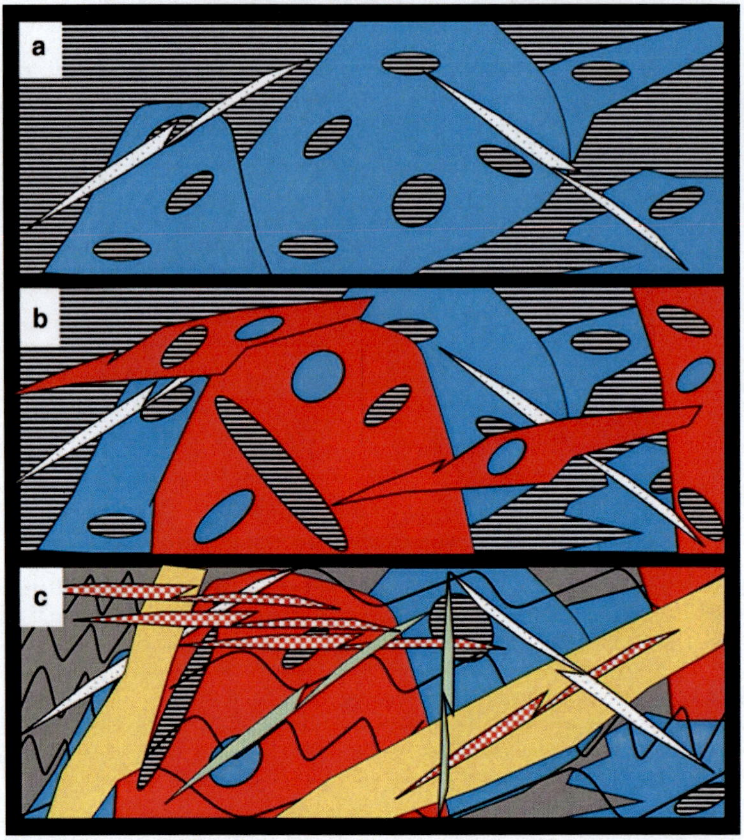

Fig. 3.28 Geo-intrusive-deformational model for the Ladakh batholith in the Leh Region. **a** Early gabbroic and dioritic plutons intrude a turbiditic sequence of sandstones siltstones and shales. The sedimentary rocks are trapped as xenoliths within, and screens between, the gabbro–diorites. Basic dykes form part of this early batholith stage. **b** Granodiorite–granite plutons, sills, dykes, veins, and other minor intrusions intrude the gabbro–dioritic–schistose basement sequence. The granites contain xenoliths of schist, gabbro–diorite, and basaltic dyke material. **c** The batholith is affected by the main phase of deformation and is sheared, folded, thrusted, and faulted depending on crustal depths. Shears (yellow) cut and deform the basement, producing strong foliations. Basalt, Andesitic Quartz–Porphyry, and Rhyolite–Granite intrude as minor intrusions and granitic sill complexes. Many minor intrusions are syn-later deformation, with some being mostly post-deformation but affected by latest stage higher angle brittle faulting

Reference

Kumar S (2020) Schedule of mafic to hybrid magma injections into crystallizing felsic magma chambers and resultant geometry of enclaves in granites: new field and petrographic observations from Ladakh Batholith, trans-Himalaya, India. Front Earth Sci 8:551097. https://doi.org/10.3389/feart.2020.551097

Chapter 4
Structural Geology of the High Strain Zone: Ladakh Batholith, Leh Region, NW India

Abstract This chapter focuses on a NW–SE trending high strain zone (HSZ) within the southern part of the Ladakh batholith, extending some 40 km from Taroo in the NW to Stakmo in the SE. Results from detailed field observations and structural measurements accompanied by detailed analysis of photographs and photo-mosaics are presented. The HSZ is characterised by a high density of contiguously-fractured plutonics, an abundance of strongly foliated rocks, and high densities of brittle–ductile shear zones. These structures strongly influence the shape and overall geomorphic form of local topography producing isolated thrust-escarpment-ridges, and well-featured contiguous ridges. Thrust structures are present at a range of scales from micro to mountain. The overall thrust geometry is dendritic in form with concave-upwards branch secondary structures, rooted and asymptotic to basal sole thrusts, bending upwards towards and asymptotic to roof thrusts. This dendritic form, when mapped out clearly indicates kinematic thrust directions on the outcrop to mega-scale. Other classic thrust structural elements are present including, horses/discrete individual thrust units, ramps, pop-up structures, flower-structures, antiformal thrust stacks, and related synforms: many examples of these structures are presented. A combination of structural data and interpretations (field-outcrop, geological mapping, photographic interpretation) indicate thrust vergence to the NNE–NE and SSE–SW. In some areas the two vergence directions are pene-contemporaneous, whilst in others one direction dominates over the other and there is a clear time history to the thrust vergence. The overall regional structural architecture is presented via a series of structural cross sections and a klippen model, which is indicative of a highly fractured basement, with crustal blocks (klippen) varying in scale from c. 1–3 km^2 to 7 km × 15 km. Individual klippen moved NNW to NW, and SW with additional strike-slip motions. Kinematic indicators close to klippen boundaries are complex indicating a range of rotational and translational klippen–tectonic-interactions.

Keywords Thrusts and shears · Structural geology · Fractured geological basement

M. G. Petterson, *Himalayan Thick-Skin Basement Deformation of the Ladakh Batholith, Leh-Ladakh Region, NW India*, Springer Geology,
https://doi.org/10.1007/978-3-031-31566-4_4

4.1 The Structuctural High Strain Zone (HSZ)

This analysis of the structural geology of the Leh–Ladakh region subdivides the Ladakh batholith into two zones: (1) a southern-south-western high-strain zone and (2) a northern-north-western lower strain zone (Fig. 4.1). The high strain zone (HSZ) extends throughout the study region, from Taroo and Phyang in the NW, through the central Leh and Leh valley region, to Shey, Stakmo, Thiksey, and Stakna in the SE (Fig. 4.1). The Leh–Ladakh region High-Strain Zone (HSZ) extends some 38–40 km NW–SE, is c. 15 km wide SW–NE at its maximum extent within the Leh valley region, and narrows to 3–5 km in the Taroo valley and Stakna–Stakmo regions.

The HSZ is characterised by: (1) a relatively high density of highly-fractured rock; (2) an abundance of strongly foliated plutonic rocks; and (3) a high density of thrust zones, some with ductile shear components. Strongly foliated, fractured, and sheared rock is not ubiquitous throughout the region, but these features are common. The highest-strain zones within the HSZ exhibit high-grade rock-foliations, and/or a high frequency of penetrative fractures and/or a high density of thrust faults. Fracture spacing can be mm to dm in scale, with mm–cm scales being common. Foliation thicknesses vary from mms to c. 1 cm and can contain strongly aligned/extended xenoliths and minerals, and significant veining (quartz, tourmaline, chlorite, epidote, calcite). Shear zones occur at mm–m scales, whilst thrust structures range up to km scales.

Figure 4.2 illustrates aspects of the fracturing that are such a characteristic of the HSZ. The Tsemo–Leh Palace Ridge (Leh-valley area), gabbro–diorite displays a remarkable set of strongly-oriented, high-density fractures, extending over a strike length within the photograph of c. 700 m. Note how these structures control the

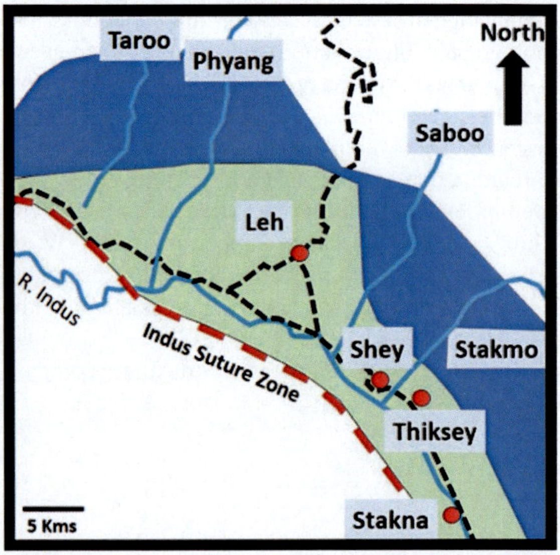

Fig. 4.1 Location of the high-strain structural zone (Green area) and low-strain zone (Blue area), Leh–Ladakh region

geomorphology of the ridge, forming strong escarpment and bench features. The lower photograph in Fig. 4.2 presents a fracture network with a spacing interval of c. 0.5–1 m, strong preferred fracture-orientations, and strong lateral continuity. A thrust-fault displaces the fracture network in the lower part of the photograph, demonstrating the detailed local structural complexity of outcrops within the HSZ.

Figure 4.3 presents lithological characteristics within the gabbro–diorite of the Shanti Stupa Ridge, Leh-valley area. Photographs on the left-hand side of Fig. 4.3 display highly developed gneissic foliations formed from the segregation of leucocratic quartz–feldpar dominated and melanocratic hornblende dominated mineralogies. Individual foliations are mms thick. Xenoliths, where present, are highly

Fig. 4.2 The upper photograph is taken from the Leh valley Shanti Stupa area and presents the Tsemo Ridge on which the Leh Palace (lower grey building) and upper Tsemo Gompa (upper left white and red building) are situated. Note the highly developed, high-density, large scale, strongly aligned, and lateral continuity of the fracture network (within a gabbro–diorite) which extends from the top-left to the bottom right of the photograph (c. 700 m in length). The lower diagram presents highly fractured rock below Shanti Stupa Ridge. The white monument with a golden spire is a stupa c. 3 m high. The gabbro–diorite displays a regular, fracture system, with major fractures 50 cm to > 1 m apart, strongly aligned, extending from the upper-right to lower-left of the photograph, subtending an angle of c. 70° with the photograph horizontal. A thrust fault locally displaces and re-orients the fracture geometry, forming the prominent bench on which the stupa has been built

extended within the foliation planar fabrics, with length/width ratios of up to 10:1 to 6:1. The xenoliths have a strong preferred orientation and are one of the key constituents of the foliation fabric. Foliations and xenolith long axes can define local 'S' shaped fabrics. The right-hand side photograph displays a high-density, strongly-oriented sequence of fractures extending some 8 m. The rock is intensely fractured locally, with only mm–1 cm separations between individual fractures. The fracture systems are strongly controlled by local ductile shear systems defined by S–C fabrics, and growth of minerals such as actinolite, tourmaline, chlorite, and epidote, aligned within the shear fabrics. The presence of kinematic indicators within the shear zones and the general geometry of fracture systems allows for the systematic mapping of shear/thrust directions across the HSZ.

Fig. 4.3 Three photographs present aspects of the sheared and strongly foliated character of the HSZ. The two photographs on the left (LHS) present strongly developed foliations, with foliation thicknesses of c. mm, characterised by the segregation of hornblende rich melanocratic foliations and quartz–feldspar rich leucocratic foliations, together with highly extended and aligned xenoliths. Note the ductile-highly sheared nature of the lithologies (pencil for scale in lower photograph is 15 cm long and applies for both the LHS photographs). The right-hand side photograph shows a strongly aligned fracture network extending some 5 m. The fractures are strongly influenced by the presence of numerous local anastomosing high-grade shear zones

4.2 Taroo High Strain Zone Structural Geology

The Taroo component of the HSZ is the narrowest in the region, some 5 km long NE–SW. It is characterised by spectacular thrust structures and areas of high density fracturing within a predominant granitic lithology (Fig. 4.4). There are textbook examples of the generation of strong penetrative fabrics through secondary deformation within the plutonic rock mass. Most of the structures are brittle in nature, although there are elements of brittle–ductile behaviour. The fractures are often strongly laterally continuous and once formed allow the rock mass to be deformed with strong fracture control. Fracture dip and strikes generate the deformation architecture from small to large scales. Mapping of the fractures elucidates the structural shapes and formations, allowing for detailed structural definitions and interpretations.

Figure 4.5 provides a particularly striking example of the character of the frontal part of the HSZ within the Ladakh–Leh area. A prominent escarpment ridge, detached from the main Taroo valley, here termed the South Taroo Ridge, beautifully exposes elements of a thrust-back thrust structure. The view in Fig. 4.5 is from the SE and highlights the SSE-directed thrust structures. Strong escarpment geomorphology terrain with well-formed scarp and dip slopes are controlled by fractures and lithological variations. There is a strong NNW dipping series of benches within the escarpment ridge. Detailed field and photograph analysis reveals a dense array of fractures, which are curvilinear in geometry, asymptotic in the lower part of the ridge and opening out in the higher part of the ridge. 'Tree' and 'branch' fracture geometries indicate the direction of motion of thrust structures. Smaller thrust/fracture units join larger 'master' thrust structures. Fracture dips vary from near horizontal to vertical. Very

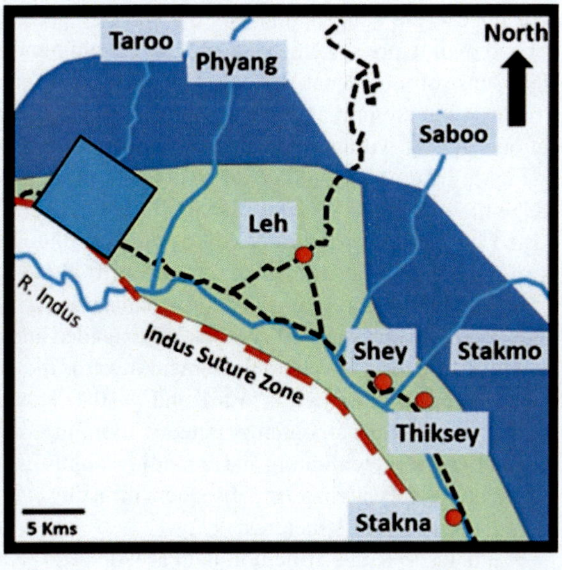

Fig. 4.4 Location of Taroo high strain zone (light blue square within the figure)

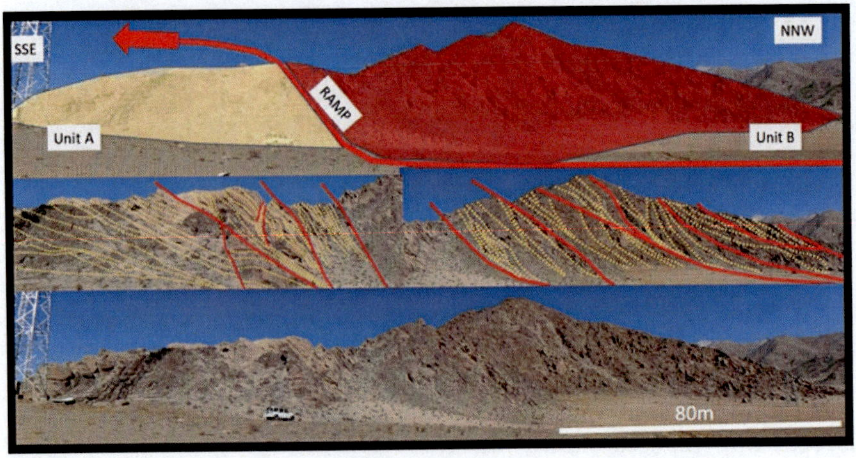

Fig. 4.5 Photomontage of the strongly featured South Taroo Ridge viewed from the SE. Lower diagram is the photographic image, the central diagram presents the main fracture trends (yellow) with the red lines indicating more major thrust structures, and the upper figure presents the larger scale interpretation with thrust units A and B and intervening ramp structure. See text for details

high density, sub-vertical to steeply dipping fracture zones characterise ramp structures. This part of the ridge is interpreted as a thrust stack with numerous individual horse structures. Fundamentally it can be divided into two units (A and B) with a ramp structure separating the two units. The ridge is part of a southwards-propagating thrust system with unit A being the youngest part of the thrust stack.

Figure 4.6 is a view of the northern part of the South Taroo Ridge. The geometries here oppose the southern part of the same ridge. Prominent escarpment benches defined by fracture systems and/or lithological boundaries dip south and indicate an NNE sense of movement. These structures are less common than the south-directed structures but are probably pene-contemporaneous, as both sets of thrust structures cut one another with no predominant cross-cutting tendency observed.

Figure 4.7 presents outcrop detail of the northward verging part of the South Taroo Ridge (part of Fig. 4.6). Four distinct lithological units can be discerned. These are hosted by granite to diorite medium grained plutonic rock with variable degrees of diorite–gabbro xenoliths. The thrust-shear zones take advantage of lithological variations within the pluton. Lithological boundaries help to focus shear and be fault processes/structures. Xenoliths are extended and aligned to the maximum local extension direction ($\sigma 3$) with movement towards the NNW. Xenolith long/short axis ratios are mostly between 4 to 5:1 and > 10:1. Kumar (2020) interprets examples such as these as pene-contemporaneous mingling of basic and acid magmas, with magmatic processes aligning and extending xenoliths. Whilst this may have occurred the structures are caught up in subsequent thrusting events that accentuate any primary co-magma mingling structures.

South Taroo Ridge structural data at whole-ridge and individual outcrop scales, together with observations from other parts of the HSZ suggest a model as

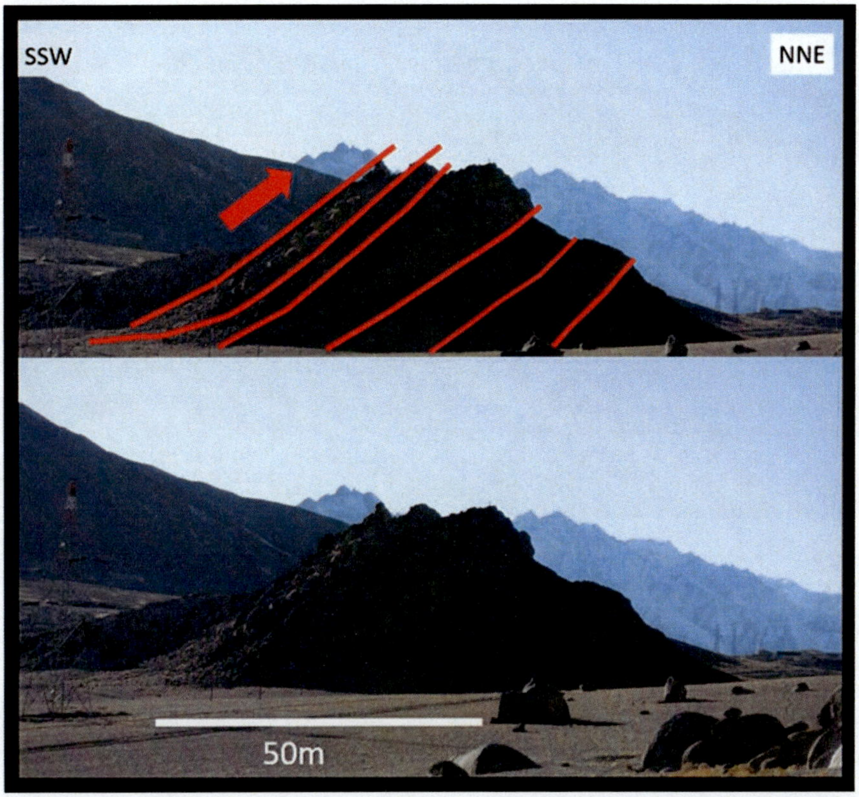

Fig. 4.6 Northern part of the South Taroo Ridge as viewed from the NE. A well-developed escarpment rises above the Indus Plain. Individual bench topographic units are defined by strong fracture systems and lithological boundaries. These benches/fractures/lithological boundaries dip southwards and have a northerly directed thrust movement. The thrusts are interpreted as approximately coeval back-thrusts to the main southwards-directed frontal thrust stack

presented in Fig. 4.8. Either opposing tectonic forces/vectors and/or a predominant tectonic vector meeting a resistant litho-tectonic obstacle led to the generation of a thrust/shear-back-thrust/shear scenario. The bulk of structures are brittle, with elements of ductile behaviour suggesting that the deformation took place within the upper part of the brittle–ductile transition zone in the South Taroo area. This convergent tectonics produces structural trends which diverge upwards in an open-concave manner from a vertical-steep central 'root-zone' or 'pop-up' structure. The overall geometry is a 'flower' structure. Lithological variations within the plutonic host reflect original lithological heterogeneity and the application of ductile stress segregating zones of variable rheology.

At the southeastern-most part of Taroo a highly fractured ridge is exposed (Figs. 4.9 and 4.10). A high-density fracture network within a granite-granodiorite pluton dips at 25–35° towards the south. The overall fracture geometry is interpreted

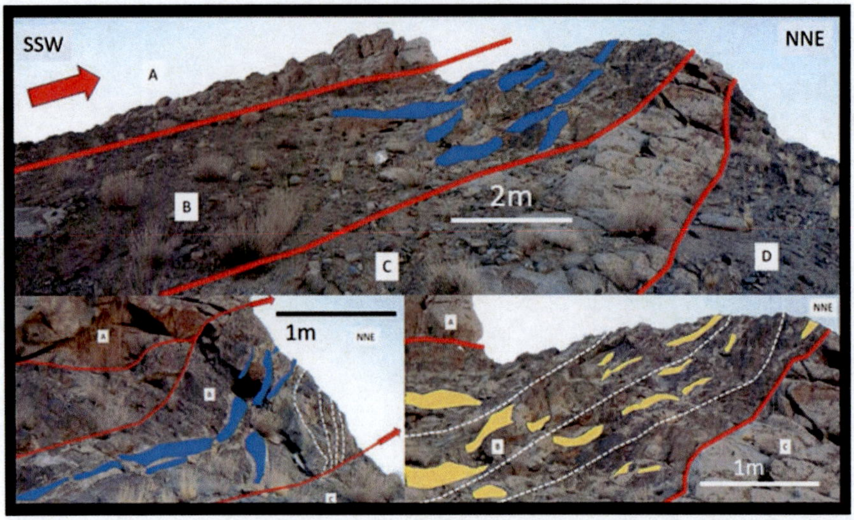

Fig. 4.7 Detailed outcrop observations from parts of the South Taroo Ridge as presented in Fig. 4.6. Note the variations in lithological units (A, B, C, D) which are granitic to dioritic host lithologies with varying abundances of gabbro xenoliths (some xenoliths are picked out in yellow and blue). The xenoliths are extended and aligned parallel to the main extensional shear direction (σ3) with long/short axis ratios of 5–10:1. Red lines are lithological boundaries which may also show evidence of displacement. Red lines define fault planes. Overall local thrust and shear movement is towards the NNE. The xenoliths may indicate intermingling of basic and acid magmas with some strain related to magmatic processes (e.g. Kumar 2020) but tectonic shear has intensified any original magmatic structures

as a north-directed pair of low-angle sole and roof thrusts (in red in Fig. 4.9) with co-eval 'secondary thrusts' meeting the sole and roof thrusts asymptotically and forming and array of southward-dipping high angle fractures in between. A central zone with fracture dips at c. 70° north complicates the overall geometry suggestive of local, south-verging back-thrusting. This situation is the reverse of the South Taroo Ridge, suggestive of penecontemporaneous north- and south-verging thrusting. Close examination of cross cutting relations suggests that the N-verging thrusts cut the south-verging thrusts (Figs. 4.9 and 4.10).

4.3 Phyang High Strain Zone Structural Geology

The Phyang HSZ zone is c. 8–10 km in length, in a north–south direction (Fig. 4.11). It is characterised by a relative high density of fracturing, shearing and thrusting.

The northernmost part of the HSZ in Phyang exhibits significant ductile deformation (Fig. 4.12). Diorite-gabbro plutons with a strong foliation and/or high density of fracturing are isoclinally folded and faulted, indicating deformation within the

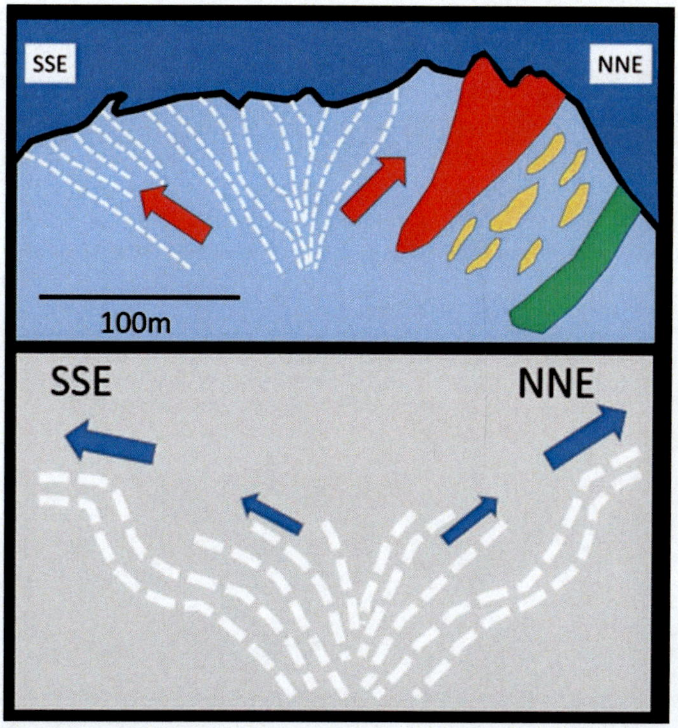

Fig. 4.8 Conceptual model of structures found within the frontal region of the Leh-Ladakh HSZ. Tectonic convergence produces divergent thrust/shear directions with structural 'pop-ups' in the central part of the structure. These 'flower structures' are present throughout the frontal, SW-most part of the HSZ, possibly reflecting A) convergent tectonic forces (SSE-NNE) or B) north/south verging tectonics meeting a strong litho-tectonic obstacle which encouraged counter-thrusting/shearing to develop. Model based on data from the South Taroo Ridge

brittle-ductile zone, though at deeper levels than the southernmost parts of the HSZ, as there is a higher expression of ductile deformation. Vergence is both to the WNW and ENE, with ENE brittle reverse faults predominating. Later granite stocks intrude.

Further south, some 2.5–3 km north of southernmost Phyang, a 1–1.5 km well-exposed ridge displays a range of structural styles (Fig. 4.13). A mixed sequence of semi-pelitic meta-greywacke schists into which granite plutons are intruded have been thrusted at 100s of m to km scales. Thrust vergence is to the NW and SE, with cross cutting relationships suggesting the pene-contemporaneity of all thrust structures. The most SE thrust unit overlies a low angle SE-dipping thrust with a ramp zone towards the NW. This thrust is cut by a SE-verging thrust unit dipping more steeply to the NW, which itself may be cut by NW-verging structures towards the most northwesterly part of the valley. A flower structure defined by a central steep zone and laterally upwards-diverging fractures is present in the NW of the valley.

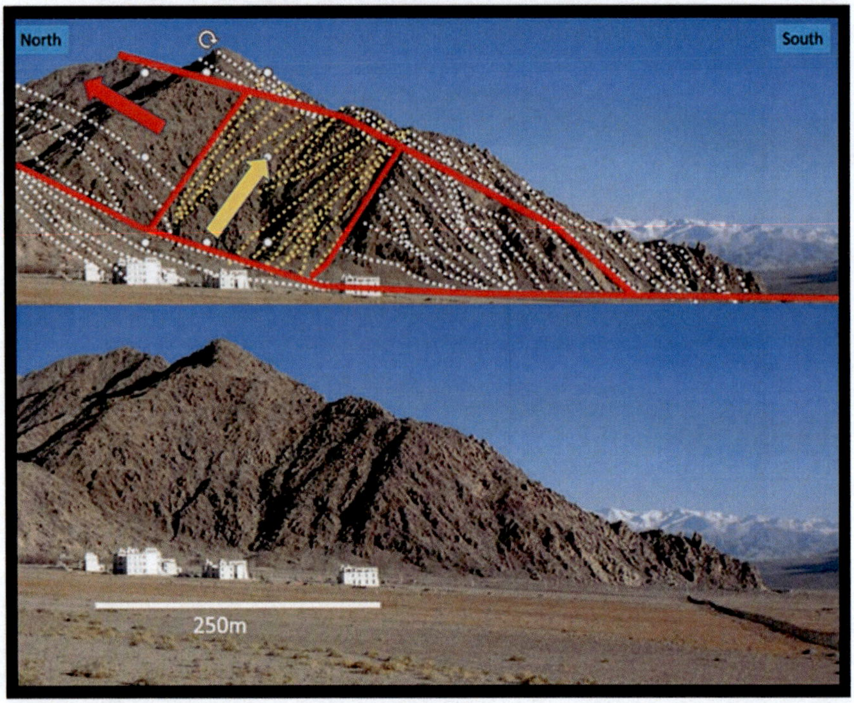

Fig. 4.9 Southeasternmost Ridge of the Taroo valley as viewed from the west. Lower part of the figure is the photograph montage with the upper part indicating southward dipping fractures in white and northward dipping fractures in yellow. The thick red lines indicate interpreted sole and roof thrust faults. The lithology is granite–granodiorite. A central ramp structure has opposing-vector south-directed steep structures indicating a thrust-back-thrust relationship

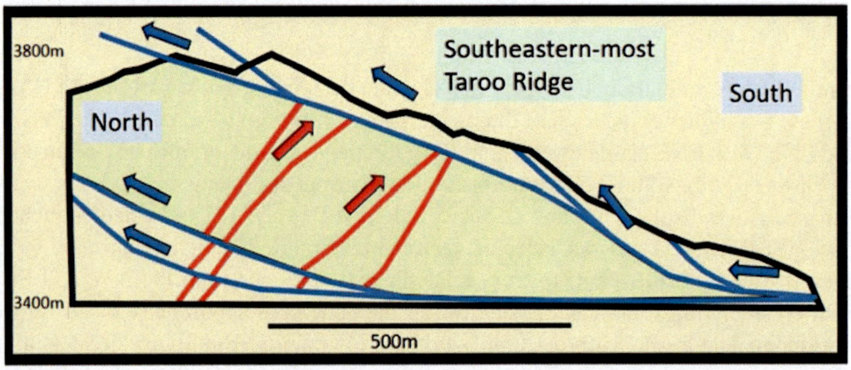

Fig. 4.10 Structural cross section based on the SE most ridge outcrop in Taroo valley. The predominant N-verging thrust structures root in a basal thrust, with an upper thrust acting as a roof thrust to the lowermost unit. Other thrusts spring from the root thrust. The N-verging thrusts cut the south-verging thrusts

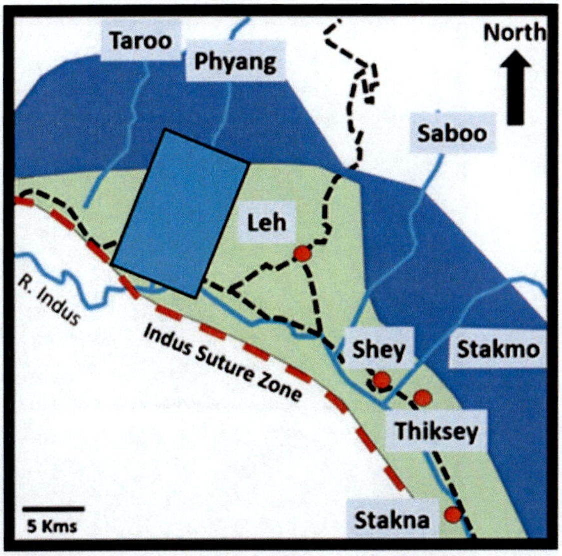

Fig. 4.11 Location of the Phyang HSZ zone (light blue rectangle)

Southernmost Phyang ridges (Figs. 4.14 and 4.15) comprise both granitic-pluton and schist-dominated lithologies. Granites are highly fractured and form well developed tor and escarpment-bench topographies. Fracture densities vary from mm–cm–dm in highly fractured zones to metre-5 m wide fracture spacings in less fractured zones. Thrust vergence is towards both the south and the north, with northern-verging thrusts being the latest in the sequence locally.

Figure 4.16 summarises the key points about the structure of the southern part of the Phyang HSZ. This section takes account of data over a c. 900 m vertical interval and 5 km horizontal (north–south) distance. The southernmost sections are along strike from one another, with the sections being some 3 km apart along an ENE–WSW strike distance. Thrust structures are 100s of m to km in scale, and up to 7–800 m in vertical height. Thrusts verge to the NW and SE, but most outcrops suggest that although the structures are in part pene-contemporaneous, the northern-verging structures are the youngest.

4.4 Leh Valley High Strain Zone

The HSZ is particularly well developed and extensive within the Leh-Town, Leh-valley area. This area extends some 15 km NE–SW and includes the areas described below as Sankar, Shanti Stupa, Tsemo, Leh Town, and Agling-Leh Golf Course (Fig. 4.17). The key structural features of the region are described using showcase outcrop examples from north–south below.

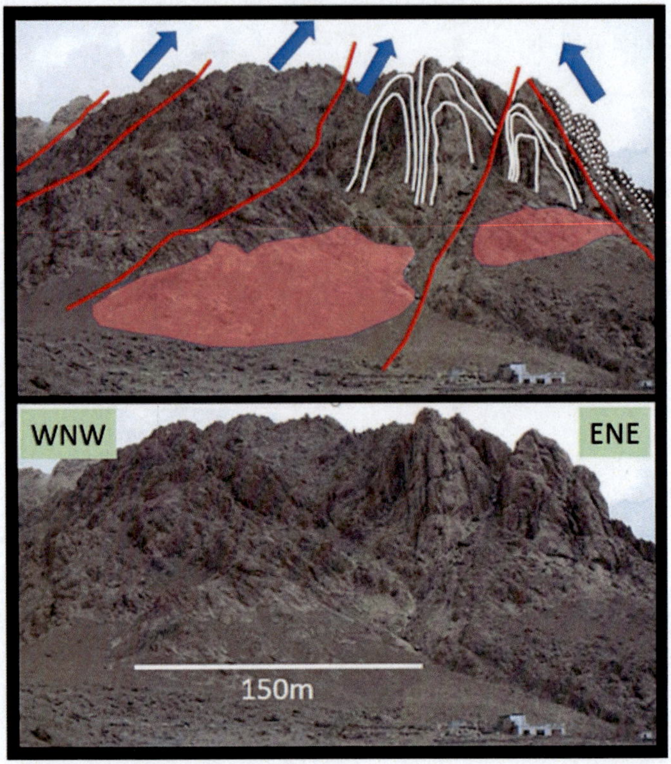

Fig. 4.12 Northernmost part of the Phyang HSZ. The high degree of fracturing within a diorite pluton facilitates brittle and ductile deformation. Tight, overturned isoclinal folds with steeply ENE dipping axial planes and related steep brittle structures indicate a degree of WNW verging thrusting/shear within a predominant ENE part-antiformal structure. Outlined red indicates later granitic pluton intrusions

Figure 4.18 is a photomontage from the northern Sankar valley region. The local geology is dominated by turbiditic meta-greywacke semi-pelite schists into which are intruded gabbro-diorite plutons and intrusive sheets. The ridge topography is strongly featured, dominated by escarpment and bench geomorphologies, with topographic benches dipping systematically northwards at c. 30°. The topography reflects schistose foliations within the metamorphic rocks and fracture-foliations within the plutonic igneous rocks. Gneissic metamorphic foliations are present within some of the meta-igneous rocks. The monoclinal, northward dipping topography and foliation dips reflect the presence of a southerly-propagating thrust stack. At least six individual thrust units separated by faults (horse structures) are visible: each of these units has a broadly increasing dip from south to north, with the youngest (southern-most) horse unit displaying the gentlest dips.

Fig. 4.13 A side valley, peripheral to the main Phyang valley, SW Phyang, exhibits a high degree of fracturing. The lithology comprises granite plutons and meta-greywacke schists. Thrust units are 100s of m long in scale, verging to the NW and SE, with cross cutting relationships suggesting contemporaneous thrust development. Upward diverging flower structures are present on the LHS of the photomontage. Granitic lithologies indicated by the pink colouring

Fig. 4.14 Southwestern-most ridge of the Phyang valley. Note how the granite weathers into characteristic tor geomorphologies. Intense fracturing forms bench and escarpment features. Fracture geometries indicate predominant earlier south verging thrusts cut by later north-verging thrusts

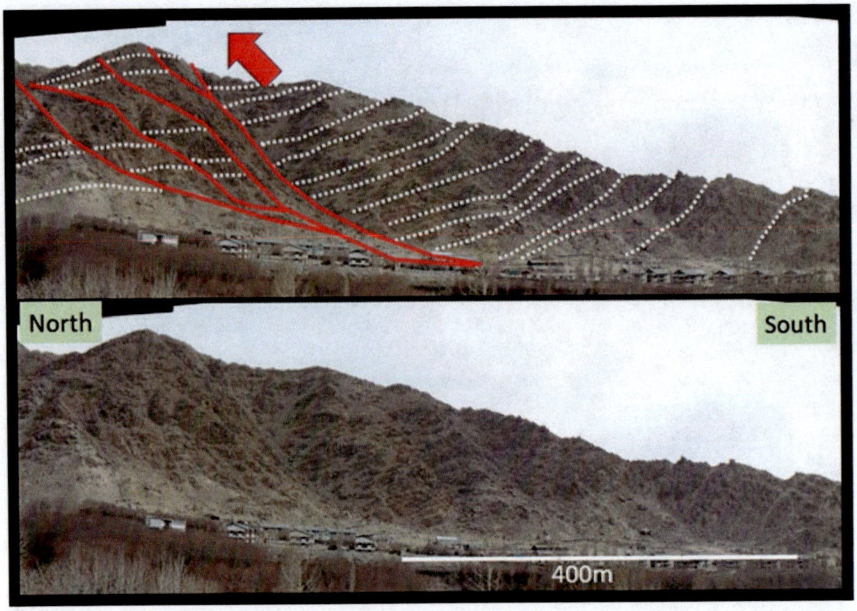

Fig. 4.15 South-eastern-most Phyang Ridge comprised of schists. Flat to gentle northward-dipping schistose layers are cut by a north-verging thrust system

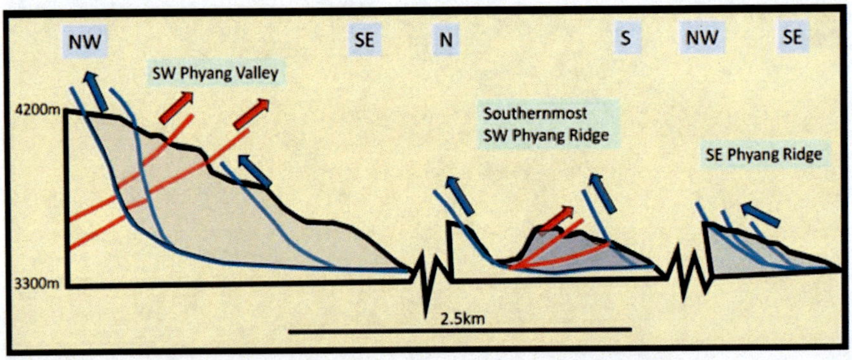

Fig. 4.16 Series of structural sections through the southern part of the Phyang HSZ. Thrust structures are on a 100s of m to km scale. N–NW and S–SE verging thrusts are present. There is a degree of pene-contemporaneity for both thrust-vergence systems, although in most outcrops NW verging structures are the youngest

Figure 4.19 displays photomontages of classic outcrops from the HSZ zone. Here outcrops at the cm to 10–30 m scale display a wide variety of highly fractured granodiorite–diorite characteristics. Some highly sheared zones are intensely fractured with a fracture spacing of mm to cm scale: Fig. 4.19 presents visible fracture densities within the photo-montage of c. 15 fractures/metre. On closer observation the

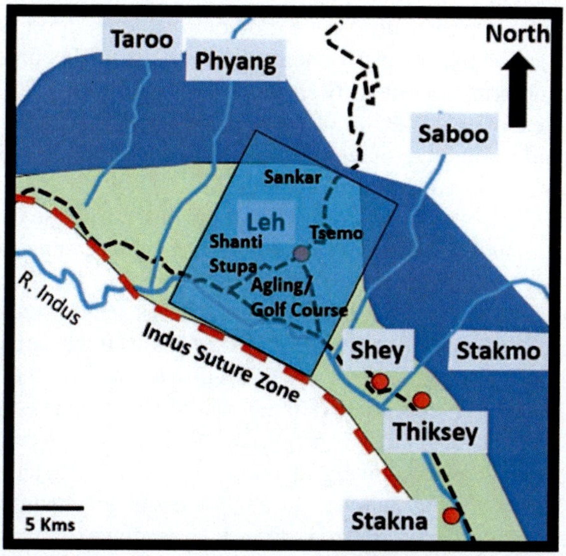

Fig. 4.17 Leh-valley region of HSZ. *Note* positions of Sankar, Tsemo, Shanti Stupa, and Agling-Golf Course

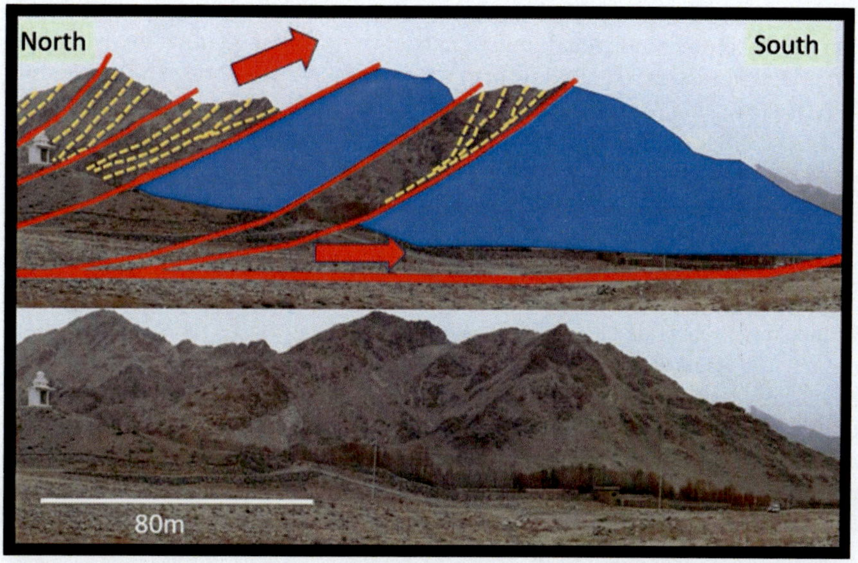

Fig. 4.18 The ridges of North Sankar valley are well featured and comprise a schist–gabbro–diorite (blue) lithology. The escarpment features reflect strong fracture-schistose foliations. Thrust faults displace rocks southwards, with individual horse-structures/thrust-units dipping northwards, within the southwards-propagating thrust system

fracturing is present at mm scales. Even at outcrop scale the level of deformation varies from high-strain high-intensity within the highly fractured zones, to moderate to low levels of deformation/intensity in the lower strain areas (Fig. 4.19). North Sankar outcrops expose c. 25–30 m of vertical and 2–300 m of horizontal continuous exposure, indicating that individual thrust units/horses are 10–80 m thick and the thrust system is exposed over at least 300 m.

Fracture and foliation dips vary from 30° to sub-vertical. These outcrops expose steepened thrust units/horse structures and thrust-ramp structures. Foliation dips are both towards the SE and NW with NW-dips predominating. These dips reflect SE and NW trending thrust propagation with SE thrust trends predominating and NW-directed thrusts forming a complementary back-thrust trend. Cross-cutting relations indicate that both thrust-vector systems cut one another: the thrusting was pene-contemporaneous in this area. Individual thrust units/horses are c. 2–10 m thick: outcrops present 6–7 m of vertical, and 40 m of lateral continuity. Foliations display a general upwards-divergent and downward-convergent geometry, with thrust-fractures increasing in number upwards, and smaller fractures propagating from larger ones in a dendritic-like trunk-branch arrangement.

At the southern boundary of Sankar, close to the Shanti Stupa region (Fig. 4.20), a dramatic isolated ridge exposes a well-formed series of synformal structures. The synforms are ESE-verging, recumbent structures, with a half-wavelength of c. 100 m. They are formed within individual thrust units some 10–120 m thick. Differential movement between individual thrust faults creates ductile deformation within the sandwiched thrust stacks, resulting in fold structures. The whole ridge is formed from an ESE-propagating thrust stack, which also cut earlier WNW-trending back-thrusts.

Thrust structures occur at all scales within the region. Figure 4.21 is from the immediate vicinity of the Upper Gompa level at Shanti Stupa. The 5 m × 0.7 m outcrop displays ramp and flat structures and excellent examples of thrust-stacking within the central zone. Local thrust directions here are predominantly towards the NE, with evidence of counter-back-thrust structures.

Figure 4.22 presents a series of views of the central-Leh, Leh Palace–Tsemo Gompa ridge. This ridge is oriented c. 1 km NE–SW and the Tsemo–Gompa summit is some 70–80 m above the valley below. Figure 4.22 illustrates a number of key points concerning the structural geology of the HSZ: (1) the region is dominated by thrust stack systems that mould the local geomorphology, often producing strongly-featured escarpment ridges, many of which form detached, isolated features; (2) laterally continuous (10s–100s–1000s of m), regular, dense fracture systems are common and a central to the geomorphological terrain architecture and to the propagation of thrust-shear systems; (3) Visible thrust-shear systems display specific thrust system elements including flats and ramps, sole and roof structures, horse/individual thrust units bound by faults, upward-diverging and downward-converging fracture geometries with dendritic trunk and branch structures; (4) the presence of systematic fracture geometries that allows the determination of thrust and shear kinematics; and, (5) brittle-dominated meso-megascale processes with local evidence for ductile deformation. The ridge is interpreted at the simplest level through a 2-klippen thrust stack model (cartoon inset in Fig. 4.22). Outcrops close to the summit/Tsemo Gompa have

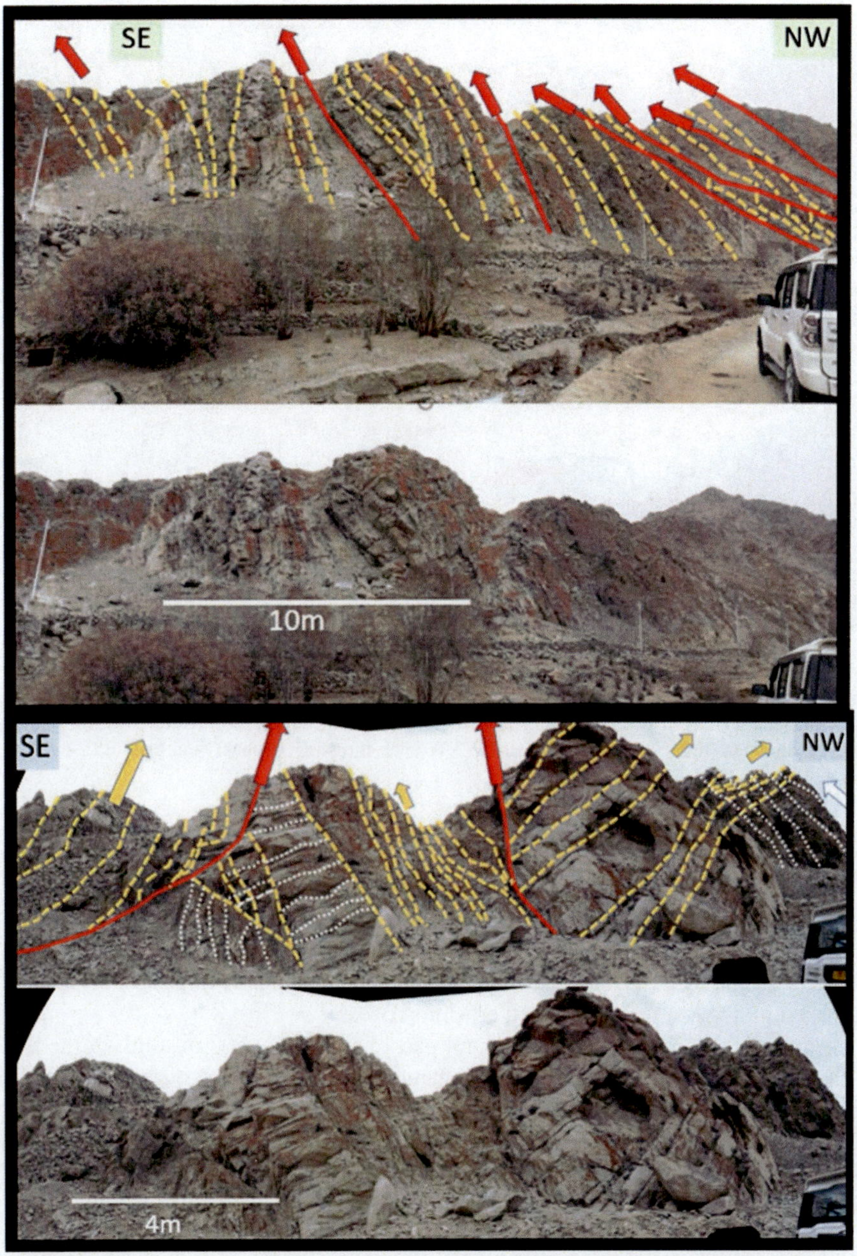

Fig. 4.19 Central-South Sankar valley outcrops are dominated by highly fractured granodiorite–diorite. Fracture densities vary but can be particularly high within the highest strain zones with a fracture spacing of mm–cm. Thrusting is predominantly towards the SE with NW-directed back-thrusts both being cut by, and themselves cutting SE-thrusts

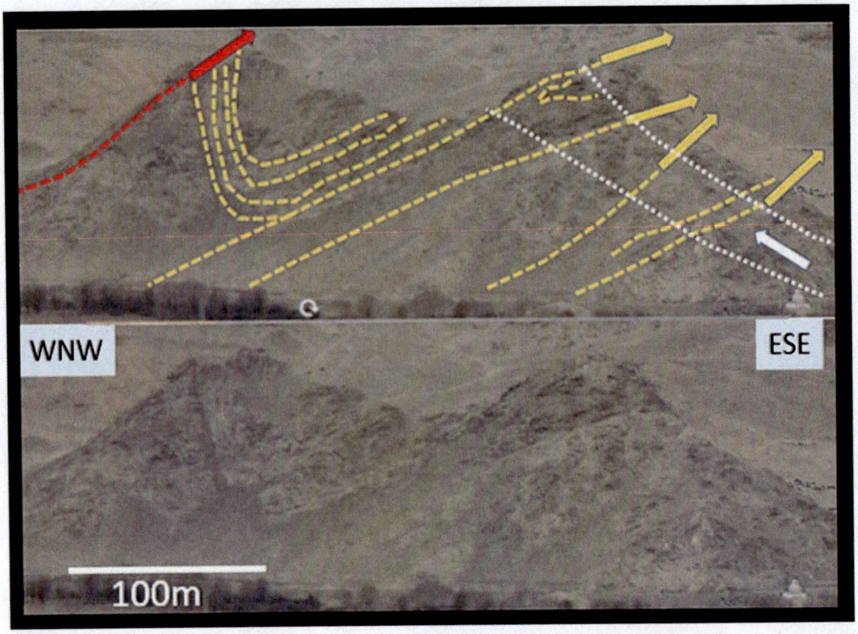

Fig. 4.20 South-Sankar. A 350 m long, 100 m high ridge exposed immediately NE of Shanti Stupa on the southern boundary of Sankar, presents a synformal structure within a deformed granodiorite. Ductile, recumbent ESE-verging synformal structures are formed through the differential movement of ESE-directed thrust faults. Evidence of WNW back-thrusting, probably cut by 2 ESE directed structures is visible

a different (SW) vergence to the lower, much thicker (Leh Palace) and predominant NE-verging klippen. Local metre scale outcrop evidence complements the larger scale observations but also reveals strike slip and counter-vergence ductile structures with tectonic 'fish' providing local kinematic evidence (Fig. 4.22 (5)). The Leh Palace–Tsemo Ridge is one of several examples of topographic ridge outliers formed as a result of thrust mechanics and erosion processes.

Figure 4.23 is from the south-central part of Leh town, 500 m south of the Leh Palace–Tsemo ridge. An adjacent taxi-stand and graveyard are situated next to the ridge. This photographic montage presents a c. 50–100 m horizontal distance, and 15–20 m high vertical exposure through a highly fractured and gneissic foliated granodiorite with consistent thrust vergence towards the NNW. Yellow dashed lines indicate local and red continuous lines larger thrust and fracture structures (Fig. 4.23).

At the southeastern-most extreme of the Leh valley area a series of ridges rise to c. 80 m above the local valley and extend over a length of several kilometres. Like many ridges in this region they form strongly-featured topography, with escarpment, bench, scarp and dip slope topographic elements. Fracture analysis from photomontages (Fig. 4.24) reveal SSE and NNW vergent thrust structures, with the predominant SSE verging thrusts cutting the NNW verging backthrusts. This is a similar structural setting and situation to the South Taroo Ridge, some 15 km NW of this location.

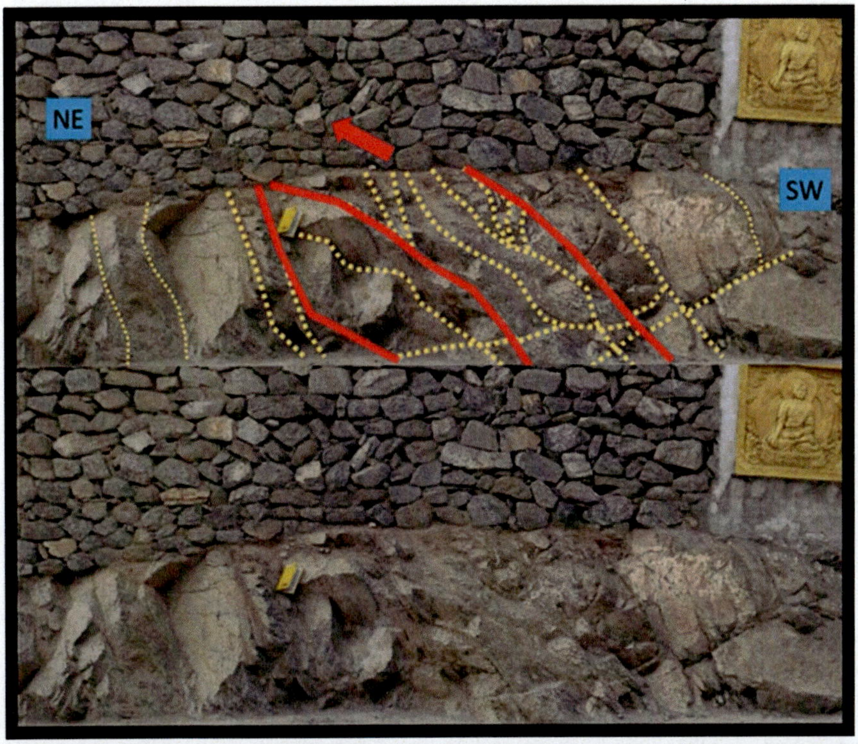

Fig. 4.21 Shanti Stupa, Upper Gompa level. See yellow notebook (30 cm) for scale. This outcrop demonstrates thrust structures at a smaller outcrop scale. The central part of the outcrop is interpreted here as a ramp structure, with local NE trending thrust movements (and some evidence of SW trending back-thrusts). Red lines indicate the main faults with yellow dotted lines indicating individual smaller scale fractures/faults

Ridges such as this become basins for the deposition of air-blown sand with sand dunes forming in the lower areas and plastered against the ridge slopes.

Figure 4.25 summarises the structural interpretation for the Leh valley HSZ region. The section collates and synthesises key structural data from Sankar in the north, to Shanti Stupa, Tsemo, Leh Taxi Stand, Agling (southernmost Leh town) and the Leh Golf Course. The main theme is one of south-verging thrust klippen, which cut north-verging back-thrust systems. Local structural cross-cutting evidence suggests that both thrust systems were active at similar times. The larger klippen are over a km thick and extend several kilometres. The same structural style scales down to outcrop, sub-outcrop, and decimetre scales. Southernmost extreme localities in the Taroo and Leh valley regions (and is described below from Stakna) have particularly strong main thrust-back thrust structural styles. In Leh valley the northern region too, around Sankar, exhibits similar main thrust-back thrust styles, but with strongly predominant south-verging thrust systems.

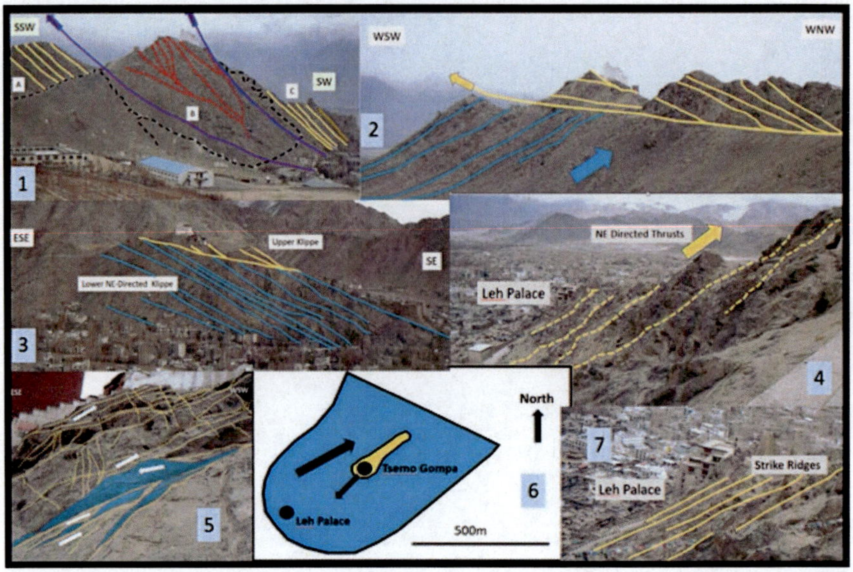

Fig. 4.22 Tsemo Gompa and Leh palace, Leh Town. One of the prominent ridges within Leh Town is the substrate to the historic Leh Palace and overlying Tsemo Gompa [Gompa/temple at the ridge-top in (1–3) above]. The ridge stands 70–80 m above the local low ground and extends c. 1 km across. This ridge is a thrust-stack ridge with prominent and contiguous fracture-cleavage-foliation penetrative fabric systems. Photographs (1–3) present a structural interpretation from different view-points illustrating the complex yet consistent fracture-thrust geometry [(A–C) within Fig. 4.22(1) marks individual thrust units]. The inset cartoon (6) provides a plan-view model structural setting with a SW-directed thin upper klippen (yellow) forming the summit, cutting a NE-directed lower thick klippen (blue) forming the bulk of the ridge. Figures (4), (5), and (7) present local views of fractures, thrusts and shears on smaller outcrop scales

Fig. 4.23 A prominent ridge in lower-central Leh town exhibits highly-deformed granodiorite with strong local gneissic foliations and a high fracture density. Thrust movement are consistently towards the NNW

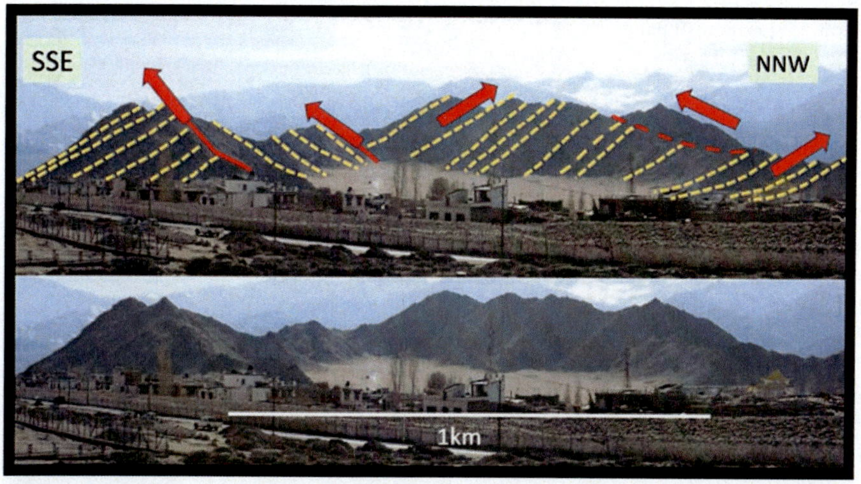

Fig. 4.24 Leh Golf Course photomontage. Ridges at the extreme southeast of the Leh valley region are strongly featured and interpreted as thrust stacks verging towards the SSE and NNW. Here the dominant SSE thrusts cut the NNW-verging back-thrusts

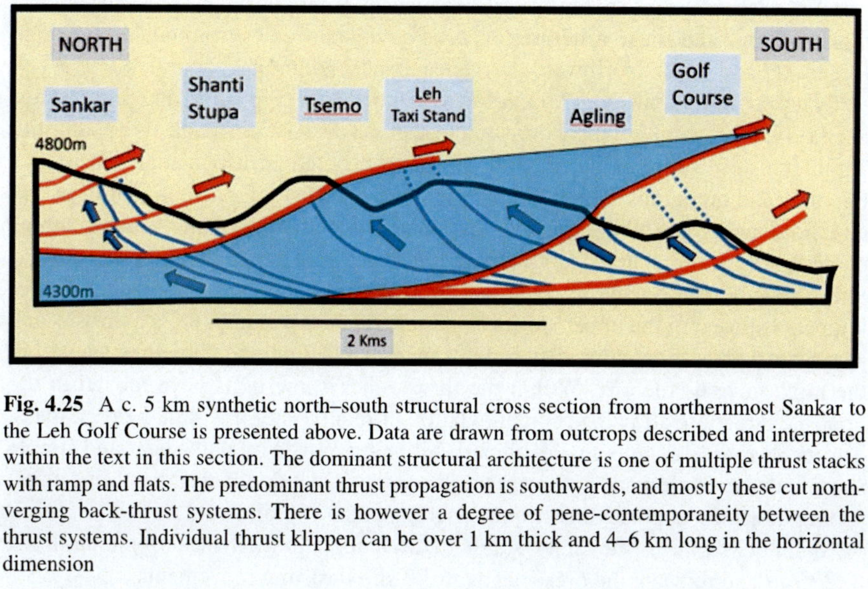

Fig. 4.25 A c. 5 km synthetic north–south structural cross section from northernmost Sankar to the Leh Golf Course is presented above. Data are drawn from outcrops described and interpreted within the text in this section. The dominant structural architecture is one of multiple thrust stacks with ramp and flats. The predominant thrust propagation is southwards, and mostly these cut north-verging back-thrust systems. There is however a degree of pene-contemporaneity between the thrust systems. Individual thrust klippen can be over 1 km thick and 4–6 km long in the horizontal dimension

4.5 Saboo–Shey–Thiksey–Stakna High Strain Zone

The HSZ area from Saboo valley to Stakna (Fig. 4.26) is some 5–8 km wide and 20 km long. The theme of high density of fracturing, brittle and ductile deformation structures, and thrust-back-thrust systems continues here.

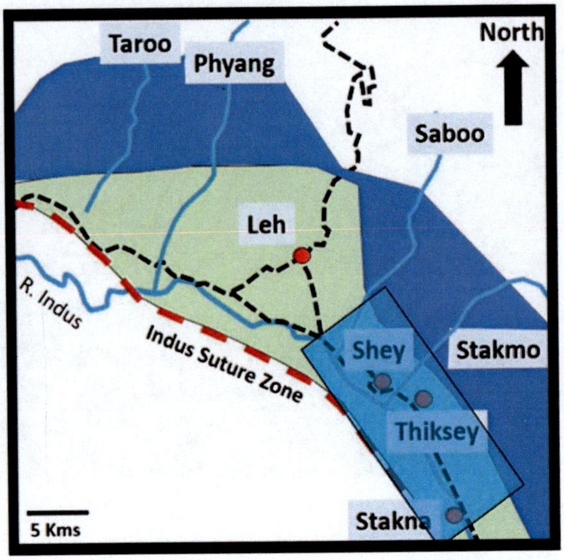

Fig. 4.26 Location of the Saboo–Shey–Thiksey–Stakna HSZ area

Lower Saboo consists of host schists intruded by granodiorite–granite and gabbro–diorite. Thrust and shear structures occur at a range of scales from outcrop to ridge (Figs. 4.27 and 4.28). Northwards vergence is most common.

Outcrops at Sindhu Ghat (2.5 km NW of Shey Palace, Fig. 4.29) and Shey Palace itself (Fig. 4.30) exhibit typical thrusting features at the southern edge of the batholith within schistose and granodioritic lithologies. Both outcrops display high densities of fractures and numerous steeply dipping individual thrust units. Dominant thrusting directions are W–WNW (Sindhu Ghat) and SW (Shey). Sindhu Ghat contains numerous back-thrust units. Both thrust directions were pene-contemporaneous. At Shey the bulk of the outcrop constitutes a lower klippen which undercuts a higher klippen, exposed in the upper slopes (Fig. 4.30). The lower klippen is a thrust stack aggregate with a thrust ramp exposed c. 4 m high by 30 m long. Dips steepen within the ramp towards the SW. Within the thrust stack some units have moved in the opposite direction, allowing accommodation of relative strain within the rock mass. This behaviour is akin to a pack of cards all being moved by pure shear and simple shear with the whole pack moving in one direction, and individual cards (thrust units) sliding down the opposite direction to accommodate overall strain.

Sindhu Ghat thrust movements are at a high angle to the bulk of regional thrust directions and indicate the presence of strike-slip thrusting movements.

The large Thiksey Monastic Complex is built upon a 1.5 km long NE–SW trending ridge, which reaches a summit of c. 140 m above the local plain (3430 m above sea level). The Monastery itself is built on a lower summit some 40 m above the local plain and town of Thiksey. Figure 4.31 presents key features of the structural geology. The ridges are strongly featured, with the dip and strike of topographic features controlled by fractures, shears, and foliations, all of which produce laterally

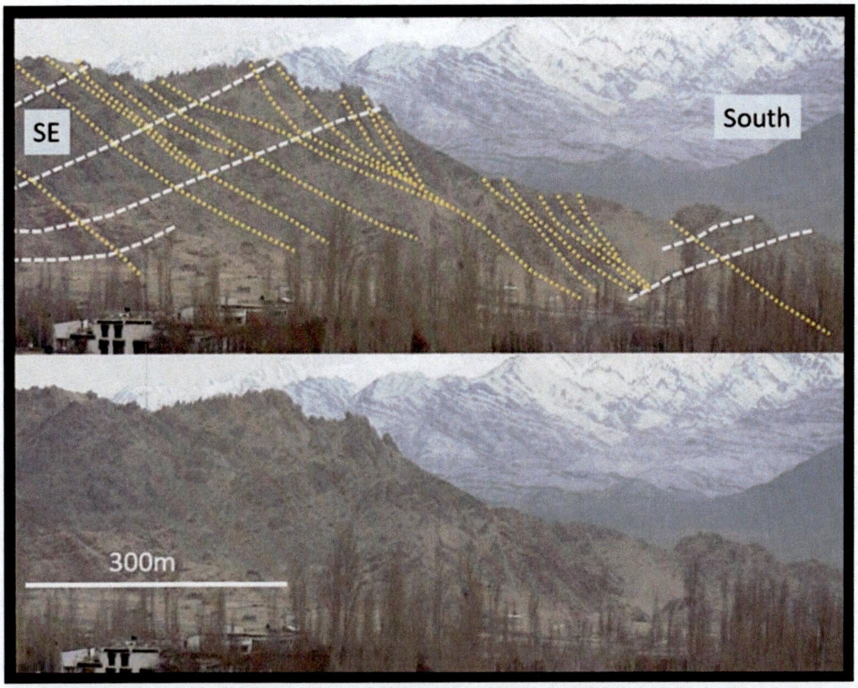

Fig. 4.27 Looking SE from lower Saboo valley. Two dominant fracture-foliation structures are present dipping south and SE. South dipping structures and most prominent and form typical thrust geometry profiles (yellow hatched lines). SE dipping structures indicate the action of back-thrusts. Predominant thrust motion is northwards

contiguous penetrative fabrics. The geology is dominated by gabbro–dioritic plutonic meta-igneous rocks with local meta-sedimentary schists. The uppermost Monastery (upper photographs within Fig. 4.31) is built on a summit below which are three strong topographic ledges. These ledges are expressions of a NE-directed thrust stack, with individual thrust units measuring thicknesses of c. 10 m. A prominent SW verging ridge breaks-through a ridge top col 150 m NW of the Monastery suggesting that it may be the youngest structure in the local area. Other SW-verging thrust related upwardly-diverging structures open towards the SW, probably cutting the NE-verging structures. Individual thrust units are 100–100s of m long.

The lower photomontage in Fig. 4.31 is from the Monastery Summit area itself looking NE down the length of the ridge. A high density of NE-dipping topographic escarpments/benches are visible. Close analysis of the geometry of these benches and related fractures reveals a thrust architecture extending over 120 m of vertical height and a kilometre horizontally. This fracture network is an interlinked southwards-SW-propagating thrust stack with individual thrusts stacked on top of one another. Thrust dip angles vary systematically from c. 10–15°/NE within the lowermost, southern-most thrust unit to 40°/NE in the highest northern-most thrust unit. Each thrust

Fig. 4.28 North of Saboo Gompa, eastern side of Saboo valley. Strongly featured ridge controlled by thrust and shear induced fracture and foliation penetrative fabrics. Ramp and flat thrust elements present. Four individual thrust units form a thrust pile with a local NNE vergence

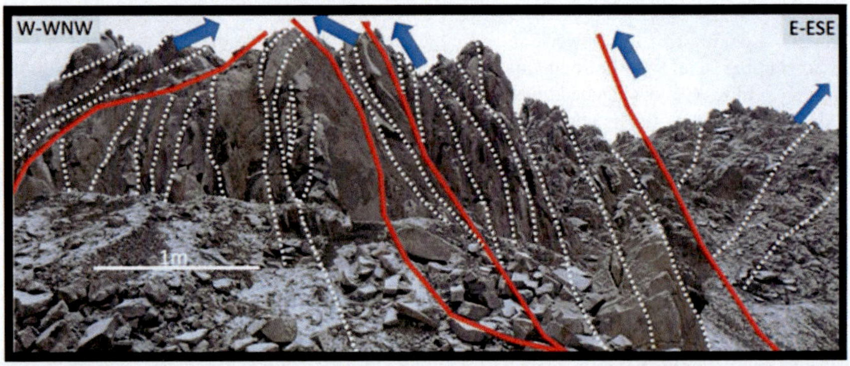

Fig. 4.29 Schist and dioritic outcrops at Sindhu Ghat, 2.5 km NW of Shey, exhibit high density fracture with steep to sub-vertical SE–S and W–WNW dips. Convergent opposite polarity vergence thrusts c. 50 cm–3 m thick meet and cross-cut each another, demonstrating contemporaneous activity

unit comprises a sole and roof thrust with dendritic upwards diverging downwards converging smaller scale structures. Individual thrust units are c. 5–10 m thick.

At the entrance to Thiksey Monastery at the termination of the ridge road are a number of highly instructive outcrops (Fig. 4.32). The overall architecture of outcrops here is one of north verging thrust and shear structures (complementary to the larger scale interpretations visible in Fig. 4.31) at 50 cm to 2–3 m scales. The outcrops here strongly exhibit both brittle and ductile structures. Brittle fractures are evident as are

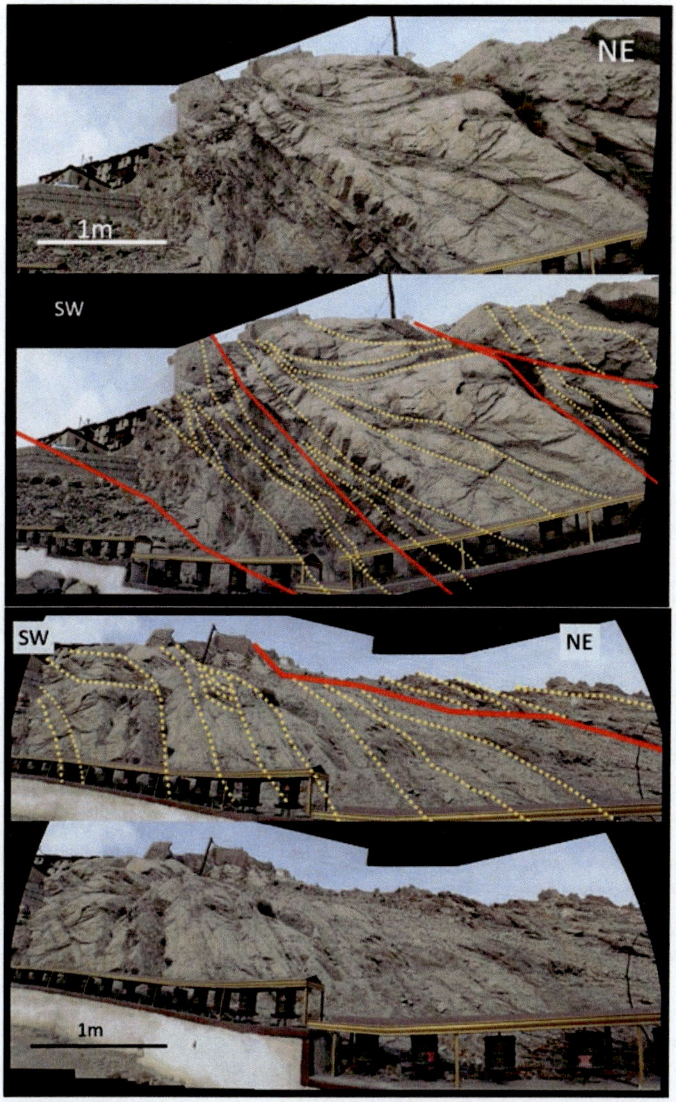

Fig. 4.30 Shey Palace is built on a granodioritic ridge. The igneous rock is strongly fractured and locally foliated. In simple terms the ridge comprises two klippen with the lower klippen occupying the bulk of the outcrop, under-thrusting below the older higher klippen. The lower klippen itself comprises a thrust pile which aggregates to form a ramp structure some 4 m high and c. 30 m long towards the left of the photomontage. Movement is towards the SW, although locally (uppermost photograph) individual units may move the opposite direction, like a pack of card's with most cards moving one direction and a few in the opposite direction

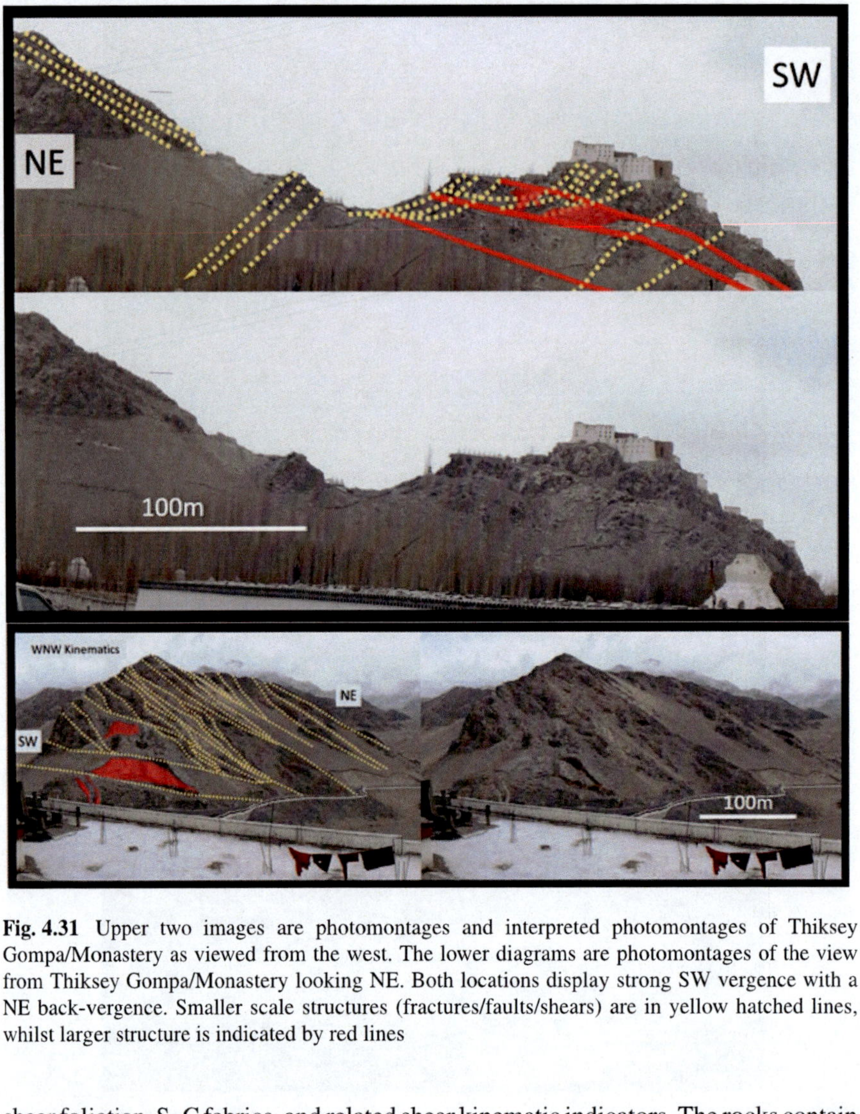

Fig. 4.31 Upper two images are photomontages and interpreted photomontages of Thiksey Gompa/Monastery as viewed from the west. The lower diagrams are photomontages of the view from Thiksey Gompa/Monastery looking NE. Both locations display strong SW vergence with a NE back-vergence. Smaller scale structures (fractures/faults/shears) are in yellow hatched lines, whilst larger structure is indicated by red lines

shear foliation, S–C fabrics, and related shear kinematic indicators. The rocks contain numerous extended lozenge-rhombohedral structures, or 'tectonic fish' typical of ductile shear zones. A particularly good example of this phenomena is presented in the lower part of Fig. 4.32. A prominent micro-granitic lithology (in white on the right-hand side photomontage and coloured blue on the interpreted photomontage) defines a well-formed rhombohedral structure indicating local shear to the top-right. Sigmoidal S-Shear fabrics within the 'fish' indicate movement to the right (south) along planar C-fabrics. These observations are confirmed by the close examination, in the field, and with high resolution, high-magnification photo-montages, by the presence of numerous other tectonic fish and S–C fabric trends. The outcrops are also highly fractured demonstrating brittle behaviour with fracture geometries indicating brittle thrust movement sympathetic to the ductile shearing. Particularly high strain has been accommodated within a comminuted mylonite zone, some 0.5–10 cm thick at the base of the lowest visible thrust unit. Field relations are suggestive of leucocratic quartz-K feldspar, high-SiO_2 granites intruding contemporaneously with thrusting and shearing activity and being themselves deformed by the deformation phase.

Stakna Gompa (Fig. 4.33) is built on a thrust ridge. The upper part of the ridge comprises a thrust stack verging SW. The thrust stack aggregates to form an antiformal thrust ramp several metres high and 10–20 m long. The dips of fractures within individual thrust units decreases from the older higher thrust unit to the lower, older thrust unit. The roof thrust zone of the lowermost thrust unit is highly comminuted with local fault breccia and mylonite lithologies produced within the fault/shear zone. Although most fabrics are brittle, ductile fabrics are locally present.

Stakna occupies a similar position within the HSZ to southernmost Saboo Outlier Ridges, the Leh Golf Course, and the South Taroo Ridge. All these locations are situated towards the southernmost boundary of the Ladakh batholith and are strongly affected by primary-back-thrust relationships. Stakna also contains outcrops that

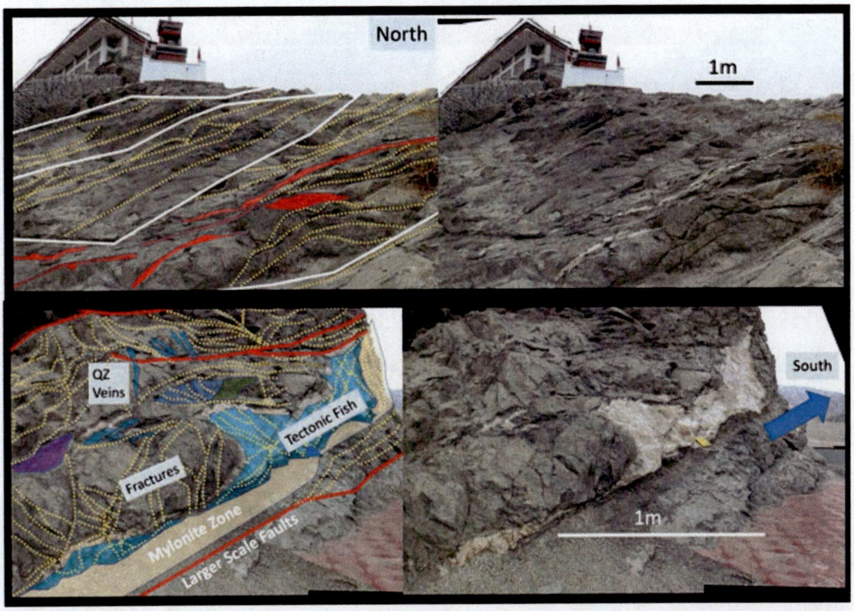

Fig. 4.32 Photomontages of thrusted and sheared rock outcrops at the entrance to Thiksey Monastery adjacent to the road termination and car park. The upper photomontages display a thrust stack verging to the north with ductile shearing of units within the thrust stack. The lower montage shows a highly sheared and thrusted rock mass with abundant rhombohedral shaped tectonic fish. These are made of both the host gabbro–diorite and later intruded high-SiO_2 granites. Different coloured lines illustrate the overall pattern of fractures and shears. See text for details

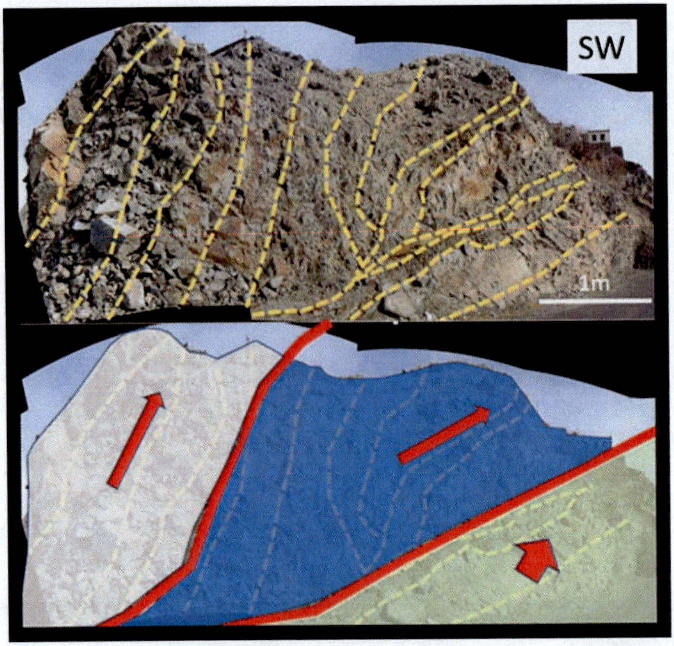

Fig. 4.33 Stakna Gompa. This outcrop displays a thrust stack, piling up to form an antiformal structure within the thrust ramp. The youngest thrust unit is the lowermost one within the photomontage

indicate contra-vergence back-thrusting as well as strike slip thrusts with kinematics subtending high angles some 70–90° discordant to the N–S trends.

Figure 4.34 presents a synthetic structural cross section of the Saboo–Shey and Thiksey Ridge areas respectively. Thrust vergence is to the north or south. In the Saboo area north-verging thrusts dominate with some south-verging back-thrusts. From Shey and further SW to Stakna SSE–SW verging structures dominate with the presence of north-verging back thrusts. Thrust structures range up to several 100s of m thick and 10 km long. In the Thiksey area late extensional faults cut all structures.

4.6 Whole-High Strain Zone Structural Synthesis

4.6.1 Primary and Secondary Fracture/Thrust/Shear Structural Data Analysis

Figure 4.35 presents raw directional data for > 60 individual shears and thrusts across the HSZ. Note the high abundance of these data for the Leh-valley sub-region, in which highly-deformed rocks crop-out over a particularly extensive geographical

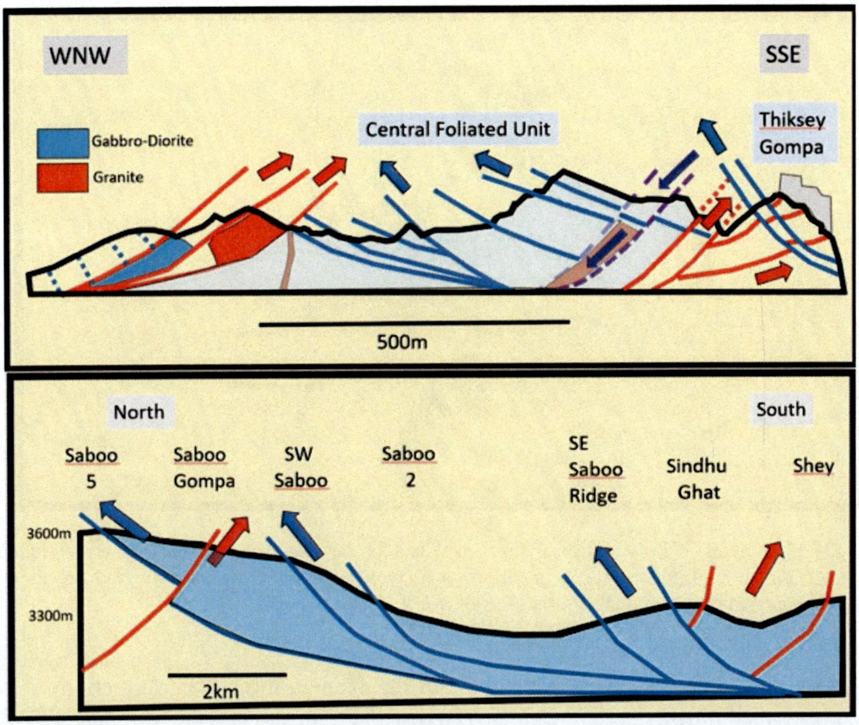

Fig. 4.34 Synthetic structural cross section accounting for data from Saboo, Sindhu Ghat and Shey (lower section) and the Thiksey Ridge area (upper section). Saboo to Shey is dominated by north-directed thrusts: these cut south-directed thrusts in Saboo but are themselves cut by S-directed thrusts in the Shey and Sindhu Ghat areas. Thiksey exhibits WNW and SSE verging thrusts which cut each other suggesting pene-contemporaneous activity. Extensional normal faults are the last structure to form in the Thiksey area. Thrust structures can scale up to 10 km long with thicknesses up to several 100s of m

area. The data reveal strong structural geographical patterns and preferred orientations. The bulk of kinematic indicators show strong NW–NNE and SSW–SE preferred orientations. Locally, there are variations caused by a range of accommodation mechanisms within the rock mass. For example, strike-slip orientations at 90° to the major thrust/shear movement locally exist, and convergent thrust/shear thrust-backthrust situations are particularly common in the southernmost parts of the HSZ. A variably complex structural situation, with a range of thrust/shear orientations, occurs in the Leh–Saboo area, and this is analysed in more detail below. Figure 4.35 shows a range in dominant kinematic thrust/shear directions from SW (and NW–SE strike slip) directions around Stakna, to SE (and NW back-thrust) directions in Taroo, northerly directions in Phyang and Saboo, and southerly directions for much of the Leh valley area.

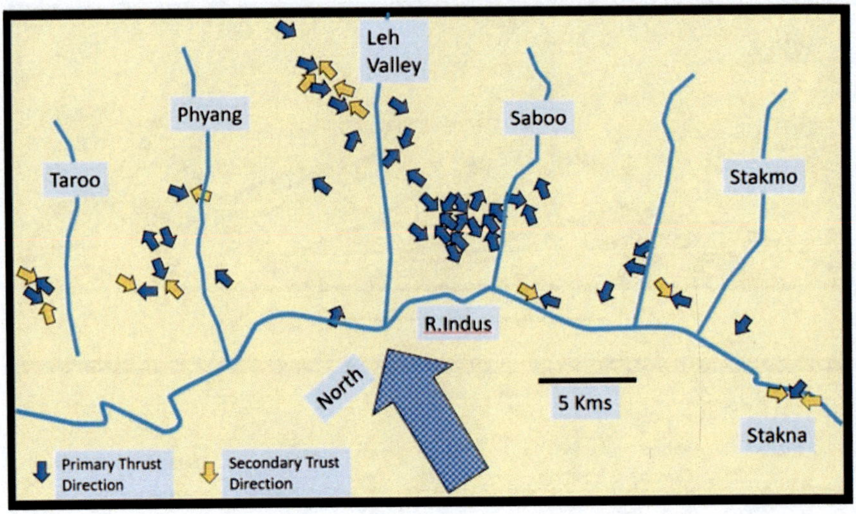

Fig. 4.35 Kinematic thrust and shear directional data for the Leh–Ladakh HSZ. Blue arrows indicate the direction of primary shears and thrusts with orange–yellow arrows showing the directions of secondary shears and thrusts (or back-thrusts)

Figure 4.36 presents directional data for the primary thrust/shear directions in a stereogram and rose diagram format. The rose diagram shows a predominant 075°–255° (ENE–WSW) orientation varying between 045°–225° (NE–SW) and 100°–280° (c. East–West). Over 90% of the data fall within the directional vector variations. A smaller data mode is between 330° and 150° (NNW–SSE) and a small data mode between NNE–SSW and WNW–SSE, which represent the strike-slip directional trends. Dips vary mostly between 10° and 35–40° with a few data points at steep 75–90° dips.

Figure 4.37 presents data for secondary thrusts/shears within the HSZ, in the form of a stereogram and accompanying rose diagram. Trends for secondary thrusts/shears are similar to primary data. They mostly represent counter-directional kinematic data to the primary thrust/shear directions, being back-thrusts and back-shears, together with strike-slip shears/thrusts.

Figure 4.38 presents the subdivision of data according to geographic sub-regions. Thrust/shear kinematic directional data varies from SW in the Stakmo/Stakna area to SE in the Taroo/Leh-valley areas, NW in the Phyang and Saboo areas, and NE for SE Phyang. Secondary back-thrusting/shearing data plot in different quadrants to the primary data: for example Taroo backthrust directions are towards the NW, whilst the primary Taroo thrust directions are towards the SE.

Figure 4.39 presents a histogram of Primary Thrust/Shear directional data. Histogram modes occur within the NNE, NE, SE–SSW, SW and NW directions. The data is summarised within the figure showing the overall north- and south-directed thrust variations.

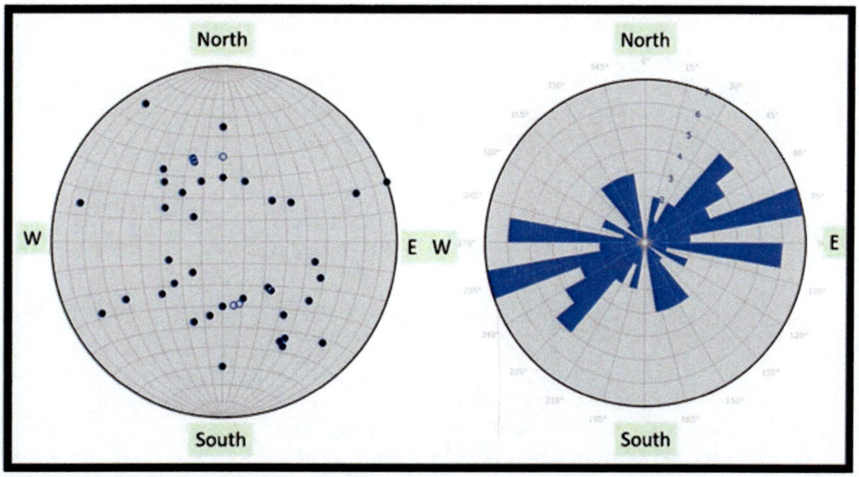

Fig. 4.36 Stereogram (poles to planes) and rose diagram (strike data) of primary thrust and shear directions for the high strain zone (HSZ), Leh–Ladakh Area. Open circles exhibit same orientational data from > 1 datum point

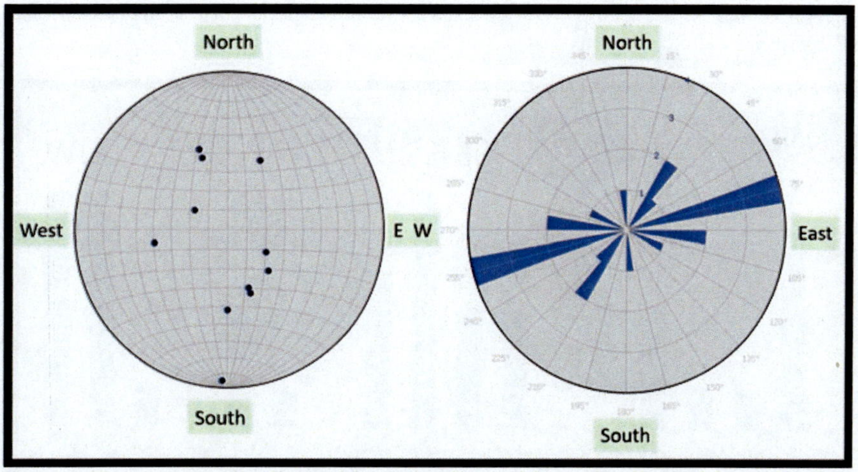

Fig. 4.37 Stereogram (poles to planes) and rose diagram (strike data) of secondary thrust and shear directions for the high strain zone (HSZ), Leh–Ladakh Area

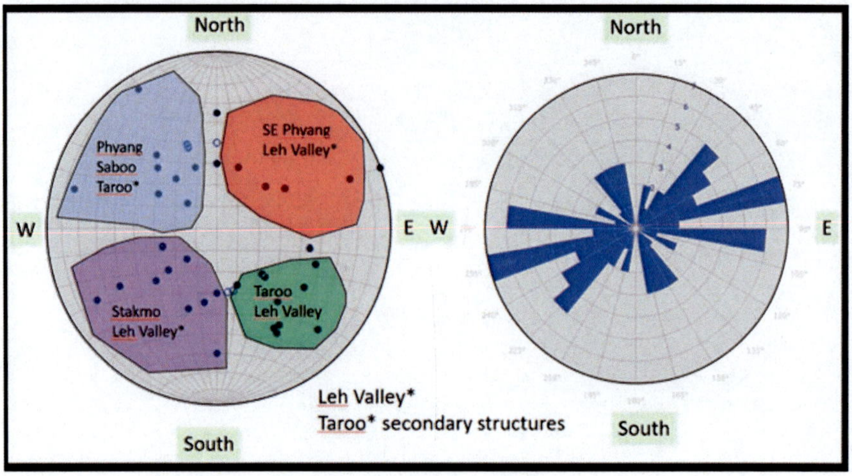

Fig. 4.38 Grouping of Thrust/Shear directional data according to sub-regions within the HSZ. Phyang, Saboo and Taroo (secondary structures) are most common in the NW quadrant of the stereogram, with SE Phyang and Leh valley secondary structures in the NE quadrant, Stakmo and Leh valley in the SW quadrant and Taroo–Leh valley in the SE quadrant. Strike data rose diagram re-presented for ease of reference. NW–NE and SW–SE thrusts/shears emanate from predominant NE–SW to E–W strike directions

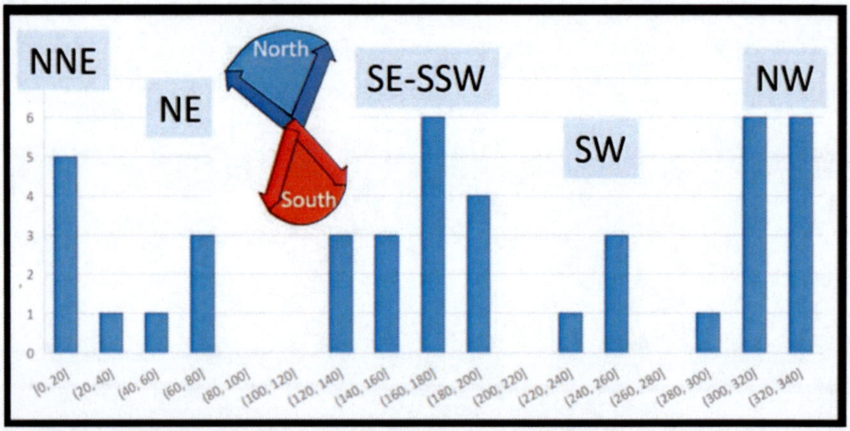

Fig. 4.39 Histogram of primary thrust/shear directional data for the Leh–Ladakh HSZ. Vectors/arrows summarise the main data trends with northward-trending shears/thrusts varying between NNE and NW and southern trending shears varying between SE–SSW

4.6.2 High Strain Zone (HSZ) Structural Data Synthesis

Figure 4.40 presents a klippen model summarising a synthetic model of the thrust/shear regional kinematic data. This model hypothesises the existence of a series of structural klippen, on scales of c. 20–200 km^2 which respond as crustal blocks to applied tectonics on a regional scale. Individual klippen are named Taroo, Phyang, Leh valley, Saboo, and Stakmo-Stakna respectively. Klippen movement directions are predominantly between S–SW and NW–NNE. Klippen interaction zones are present close to klippen-boundaries. Strongly polarised thrust movements are common within the Taroo–Phyang klippen convergence zone. The Leh–Saboo Convergence Zone highlights a particularly complex structural area close to the boundary between the Saboo and Leh valley Klippen.

The Leh–Saboo Complex Zone (Fig. 4.41) is a zone of klippen-interaction producing a wider variety of thrust/shear movements. The predominant movement is north and south with movements also NW–SE and ESE–WSW. These data indicate that locally the Saboo klippen was the stronger tectonic influence relative to the Leh valley klippen. Local Klippen interactions also facilitated strike slip rock mass accommodations as well as significant angular discordance to the predominant structural grain.

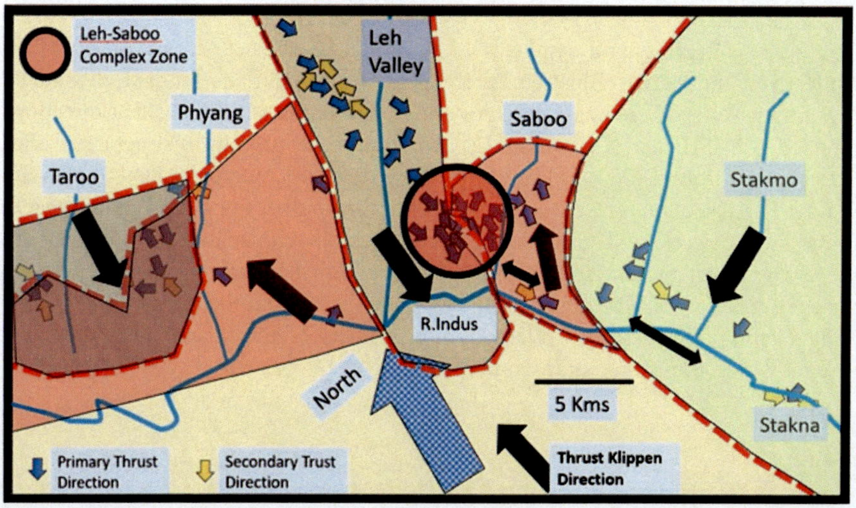

Fig. 4.40 Klippen model for the Leh–Ladakh high strain zone (HSZ). Regional thrust/shear kinematic data are interpreted here in terms of a series of klippen which have predominant movement directions towards the SW, south, NE, and NNE (large black arrows indicate klippen main movement directions). Smaller black arrows indicate the presence of local strike-slip shears/thrusts. The Leh–Saboo complex zone is highlighted. The convergence zone between the Taroo and Phyang klippen is indicated in a dark brown colour towards the LHS of the Figure. See text for further details

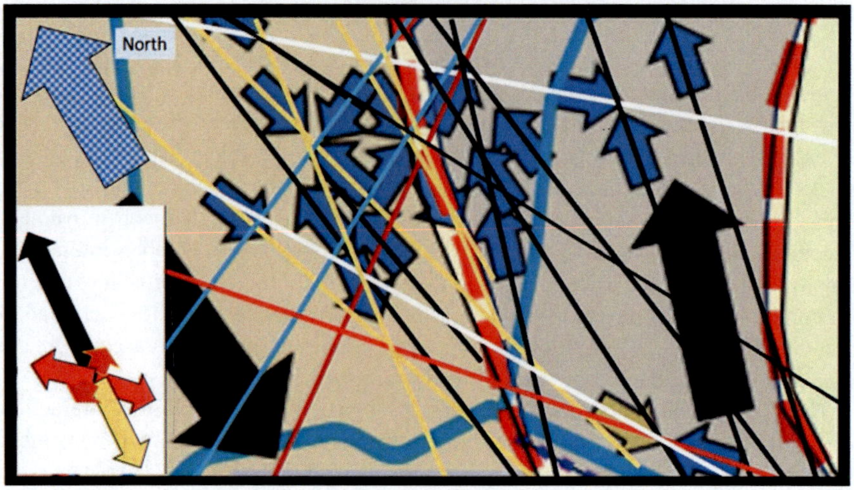

Fig. 4.41 High resolution close-up of the Leh–Saboo complex zone (see Fig. 4.40 above). The horizontal length of the full figure measures c. 5 km. Different colour lines represent the movement directions of individual thrusts (black = N–NW, yellow = S–SW, red = WNW–ESE, and brown/dark red = NE–SW). Arrows in the bottom left of the diagram summarise the key movement directions, with the length of arrow indicating the relative frequency of data

Figure 4.42 presents a synthesis of structural cross sections across the HSZ. Collectively the sections illustrate several generic characteristics of the Leh–Ladakh high strain zone. These characteristics are: (1) predominant thrust/shear directions directed SW–SE and NW–NNE; (2) the scale of thrust/shear structures varies between 2–6 km, 100s of m and 10s of m down to outcrop scales at mm-dm scales. (3) All sections present a dominant thrust/shear directions and a counter/back-thrust/shear direction; (4) Structural flower structures with central pop-up zones are present in some structural settings, particularly within the southern-most to SW-most zones of local areas. The upper right cartoon is taken from the classic South Taroo Ridge setting; (5) there is a relatively high density of thrust and shear structures present within these structural sections.

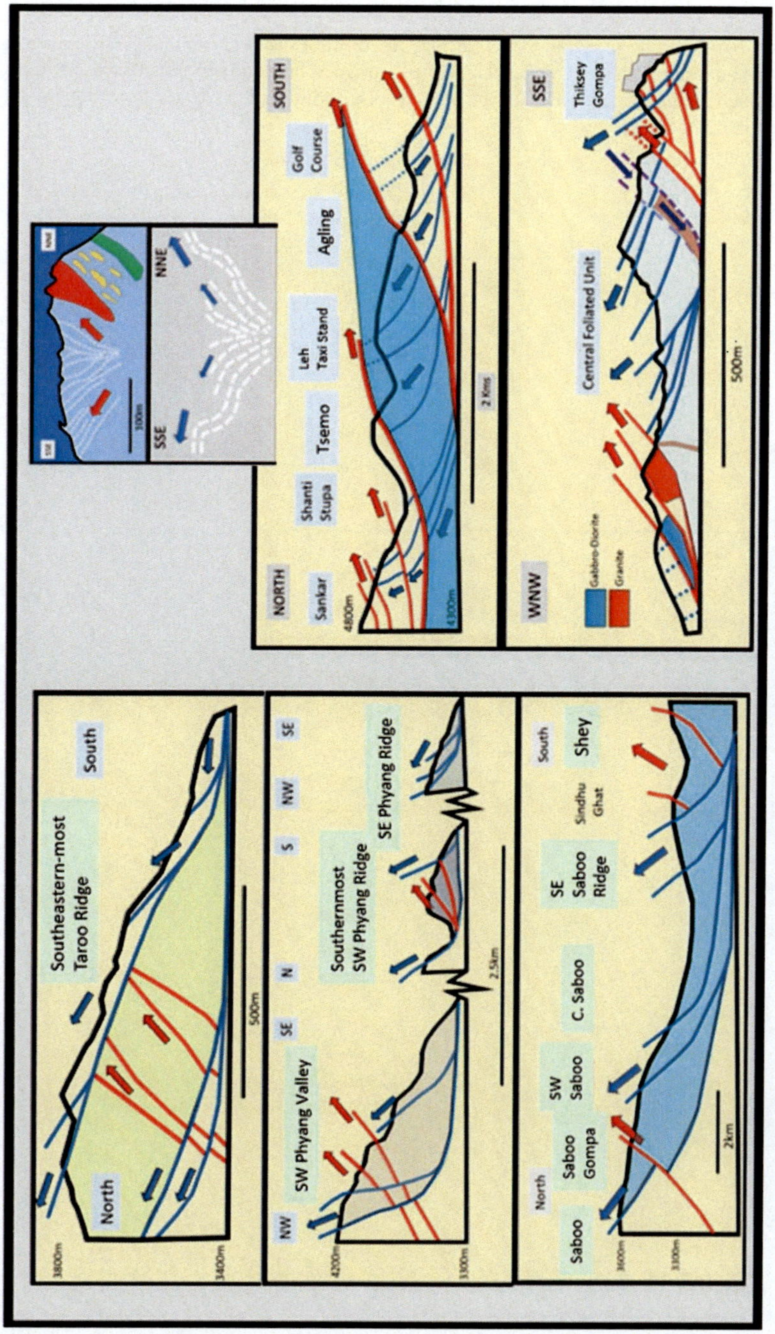

Fig. 4.42 Structural cross sections for Taroo, Phyang, Leh valley (mid-right section) Saboo–Shey, and Thiksey Gompa areas. See text for details

Reference

Kumar S (2020) Schedule of mafic to hybrid magma injections into crystallizing felsic magma chambers and resultant geometry of enclaves in granites: new field and petrographic observations from Ladakh Batholith, trans-Himalaya, India. Front Earth Sci 8:551097. https://doi.org/10.3389/feart.2020.551097

Chapter 5
Structural Geology of the Northern and North-Eastern Zone (N–NE Zone), Leh–Ladakh Region

Abstract The North–Northeastern zone of the Ladakh batholith, Leh–Ladakh Region is more heterogeneous in style, compared to the adjacent south High Strain Zone. It extends some 40 km NW–SE, and is up to 12 km wide NE–SW. This chapter reports the results of field-based observations and structural readings alongside detailed anlaysis of photographs and photomosaics. The N–NE Zone is characterised by a higher abundance of ductile deformation, which has produced a series of discrete shear zones that vary in scale from cm and metres to km wide. Shear zone-rich regions exhibit discrete zones of highly-sheared rock separated by low and lower strain zones. Other sub-zones are dominated by fold structures, particularly where meta-sediments crop out. More extensive lower strain zones are characterised by thrusts and high angle extensional faults. Shear zones produce strong gnieissic foliation, intense fracture and cleavage, extended xenoliths with long/sort axial ratios up to 20, and S–C fabrics. Brittle structures such as thrust zones have similar kinematic indicators to the High Strain Zone (NNW–NE vergence) reflecting a prevalent ENE–WSW structural grain. Shear zones mostly strike NW–SE to NNW–SSE. Fold axial planes vary in strike between NW–SE and NE–SW with the larger folds displaying a north–south axial planar strike. Structural data from the N–NNE zone support and extend the klippen-fractured basement model derived from structural analysis of the southern High Strain zone.

Keywords Shear Zones · Fractured basement · Folds thrusts and extensional faults

5.1 Introduction, Zonal Structural Styles, and Variations

Figure 5.1 shows the location of the North–North-Eastern Zone (N–NE Zone) of the Leh–Ladakh region. The area occupies the upper parts of the Taroo, Phyang, Saboo, and Stakmo valleys, and is a geographical extension of the High Strain Zone (Chap. 4). There is a gradational boundary between the two zones. The Leh–Ladakh region N–NE Zone extends some 38–40 km NW–SE, is c. 10–12 km wide SW–NE, at its maximum extent, within the Taroo, Phyang, and Saboo valleys. Unlike the

M. G. Petterson, *Himalayan Thick-Skin Basement Deformation of the Ladakh Batholith, Leh-Ladakh Region, NW India*, Springer Geology,
https://doi.org/10.1007/978-3-031-31566-4_5

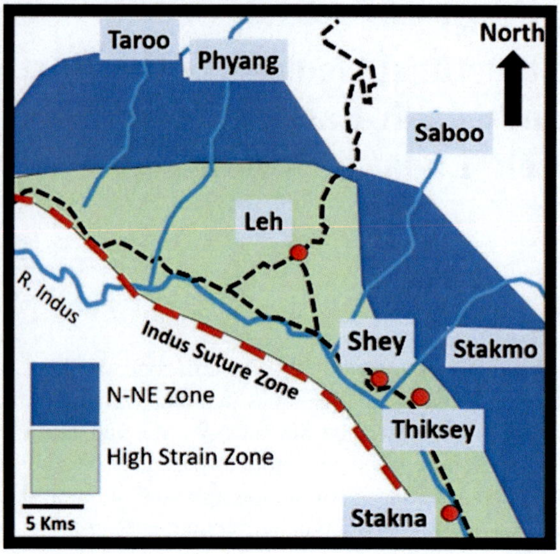

Fig. 5.1 Location of the North–North-Eastern Zone (N–NE Zone in blue)

HSZ, the N–NE Zone is more elusive to categorise, as it is more heterogeneous in terms of structural character. However, it does exhibit a range of structural settings. Key characteristics include: (1) heterogeneous strain with areas of higher and lower strain; (2) fold and ductile structures being more in evidence, compared to the HSZ; (3) intense shearing within constrained shear zones; (4) the presence of thrust zones, similar to the HSZ; (5) late extensional normal faults; and, (6) the presence of a range of deformational styles, from ductile to brittle.

This section presents an overview of the range and styles of structures present within the N–NE Zone, and will discuss examples, from NW to SE, to clearly illustrate the key characteristics of the N–NE Zone.

Figure 5.2 presents a lower strain area of the Ladakh batholith, within a side valley peripheral to the main Taroo valley, close to the narrow high strain zone in SW Taroo. The local hills rise over 200 m above the local valley floor and present a typical section through the lower-strain Ladakh batholith. The batholith itself here comprises granodiorite–granite plutonic rocks intruded by faulted Quartz–Porphyry intermediate composition dykes on the left-hand side of the image. Granitoids intrude semi-pelitic, meta-sedimentary, schistose rocks, that now form xenoliths and screens between plutons. Most of the batholith is relatively unstrained although a lower density of contiguous regular joints and cleavage fracture-foliations are present. Towards the east the granitoids are part of ESE-verging thrust zones, measuring c. > 100 to 200 m long and 50–150 m in thickness. The central and western areas of the area present an open, symmetrical, antiformal structure, with vertical cleavage present in the axial fold region. The fold axial plane strikes c. north–south. An intermediate series of Quartz–Porphyry dykes intrude the western side of the valley,

Fig. 5.2 This side valley of the main part of lower Taroo, immediately north of the HSZ is typical of the lower strain parts of the N–NE Zone. The photomosaic presents granodioritic to granitic plutonic rocks cut by a quartz–porphyry dyke (Q–P). The granitoids intrude semi-pelitic schistose host rocks which form xenoliths and screens within the batholith (MS). Granitoids exhibit a low-moderate degree of strain locally, with fracture-foliations and cleavage-fractures suggesting ESE-thrusting on the RHS of the image, and antiformal folding in the centre-left of the image

themselves faulted by ESE-striking normal and reverse faults: these are the youngest structures in the area.

In contrast to the relatively lower strain situation of Lower Taroo, Upper Taroo valley contain several shear zones measuring metres to tens of metres in width, but can span several 100s of m to a km wide of terrain: here discrete metre-10 m, or 10s of metres-wide shear zones are present with less/un sheared rock in between. Figure 5.3 is an example of a sheared area exposed over some 25–35 m in length, and 12–15 m vertically. Figure 5.3 exhibits an active slide, which slips down-dip, producing a 30° dip slope. Scarp slopes contain Glacial Till deposits and bedrock. The slip-shear plane is one of several shear planes within the rock mass. Figure 5.4 shows two metamorphic-deformational facies present within the shear zone. One facies is characterised by high densities of cleavage planes which produce closely-spaced, open, contiguous, regular, planar structures, on a range of scales along with intensely sheared rock. Higher-angle ductile fault zones cut the horizontal-gently-dipping cleavage structures. The second facies shown in Fig. 5.4 is a highly-sheared granodiorite with a discrete shear zone, 25–30 cm wide. The rock has well developed fracture-cleavage with S–C fabrics indicating top to the right movement. Some Leh–Ladakh region shear zones and thrust zones are strongly tourmalinised, with associated hydrothermal argillite clay-rich alteration (Fig. 5.5). This is evidence for shear-ductile structure assisted hydrothermal fluid activity rich in volatiles and water. This hydrothermal zone is quite extensive, stretching over 10s of m, and present to a lesser or greater degree over much of the rock mass.

Figures 5.6 and 5.7 are close-ups of plastically deformed and sheared granites. Figure 5.6 illustrates lesser and greater amounts of plastic deformation within one small area. The upper part of the photograph is granular and coarse grained. Closer observation reveals the breaking down of feldspar and quartz grains into sub-grains

Fig. 5.3 The slopes in this part of Upper Taroo experience regular landslides along shear planes. The slopes of the local mountainside are strongly influenced by a shear-dip to the NNW and have a kinematic movement sense up-dip towards the SSE. The blue colouring indicates scarp slopes of Glacial Till and bedrock left behind by the most recent slide. Direction of sliding is given by the red arrow. Car for scale (4.5 m long, 1.8 m wide)

Fig. 5.4 The LHS image presents intensely sheared rock with a cleavage–fracture density of 4–5 larger fractures and numerous smaller scale fractures per 10 cm. The shear zone is 'faulted' by ductile shear-fault systems that cut the main structures at high angles. The RHS image (compass–clinometer shown for scale by the oval) is a close-up of sheared granodiorite with S–C structures and a top to the right local sense of movement

Fig. 5.5 Some Ladakh shears and thrust zones are highly tourmalinised with accompanying clay minerals indicative of metamorphic–argillic, volatile-charged hydrothermal metamorphism. Pencil for scale (15 cm)

with strain hardening producing polygonization. There is a preferred alignment of the long axes of minerals and some segregation of dark and light minerals. The lower part of Figs. 5.6 and 5.7 show discrete shear zones 3–5 cm wide, with more intensely deformed rock displaying S–C fabrics that indicate top to the right movement.

Brittle dominated deformation zones intercalate with ductile dominant deformation zones. Figure 5.8 illustrates a thrust pile within a rock mass some 350 m long × 100 m high. Individual thrust units are 20–30 m thick, extending over the full length of the exposed granite–gabbro-meta-sediment-basalt dyke batholith. Thrust deformation has produced strong, regular, contiguous, cleavage–joint–fractures which define sole and roof thrusts, plus a dendritic smaller-scale thrust architecture. Dominant thrust motion is towards the SW with minor back-thrusting towards the NE.

Phyang valley presents both differences and similarities with Taroo, in terms of the overall structural characteristics of the valley. One commonality is the presence of shear zones and ductile structures. Figures 5.9 and 5.10 illustrate key characteristics of Phyang shear zones. The scale of the exposed shear zones varies from cm to even mms in places, to 1–5 m and 10s of m at larger scales. Plutonic rocks can be strongly foliated, forming medium-coarse grained gneisses, with segregated felsic and mafic mineral foliations. Fracture cleavage can be common.

Xenoliths are either semi-pelitic meta-sedimentary, or hornblendite in composition. The semi-pelitic ones are in a variable state of assimilation, with some xenoliths becoming 'ghost xenoliths', barely visible in outcrop. Xenoliths are moderately to highly extended with long: short axis ratios of c. 2:1 to 20:1 with an average ratio of 6–9:1. Xenoliths are mostly 10–70 cm in length. There is abundant evidence of magma: xenolith chemical interactions with the presence of mm–cm felsic or hornblenditic–biotitic reaction rims.

The central-east ridge close to the historic Phyang Gompa provides exposures of tight to open folds with wavelengths of c. 1 km and wave heights, crest to trough of c. 500 m (Fig. 5.11). To the north of the ridge fold structures are cut by several SW-verging thrust units, which themselves have thicknesses of 100–200 m. Many fold

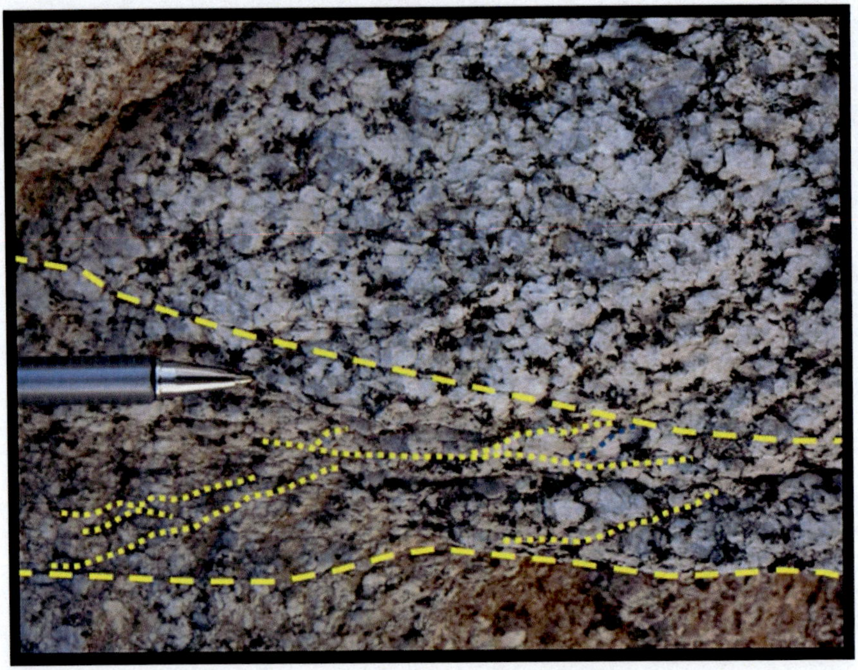

Fig. 5.6 Sheared granite from north Taroo. Upper part of the photograph shows a strong granular texture, with well-developed polygonization and grainsize reduction/recrystallisation textures in feldspars and quartz. Preferred alignment and some segregation of mafic and felsic layers is visible. The lower part of the photograph displays S–C fabrics within a more intensely sheared zone, with strong alignment of minerals, particularly mafics, and the mature development of shear-fabrics and shear-criteria. Kinematic indicators suggest top to the right. Part-pencil for scale (3 cm)

structures are disharmonic, with intra-fold faults and shears cutting, and disrupting, fold continuity. High angle faults, dipping south-westwards, are present in the central and southern parts of the ridge. Whilst originating as extensional faults, some have been subsequently reactivated as reverse faults. The ridge provides evidence for a temporal sequence of geological-deformation events interpreted here as: (1) deposition of sediments and intrusion of plutons; (2) extensional faults accommodate sedimentation, pre-deformation; (3) NE–SW folding of the central-northern ridge areas; (4) compressional tectonics produces SW-verging thrusting towards the north and NE-reverse fault motions to the south.

The northern parts of Phyang are dominated by thrusts which predominantly verge to the NW, with a limited amount of SE-back-thrusting. Thrust piles extend some 500–1000 m horizontally and 3–400 m vertically (Fig. 5.12). Thrusts mostly dip between 20° and 50° south-eastwards, with dips increasing within the thrust piles from the lower, younger thrusts to the upper, older thrusts. Sole and roof thrusts dip-angles vary from sub-horizontal to 30°. Smaller scale thrusts 'root' in larger structures allowing for the determination of vergence. Individual thrust piles vary in thickness

Fig. 5.7 Upper Taroo sheared feldspar–quartz–muscovite rich granite. Strong development of S–C fabrics with kinematic indicators showing top to the right motion. Ruler indicates mm–cm scale

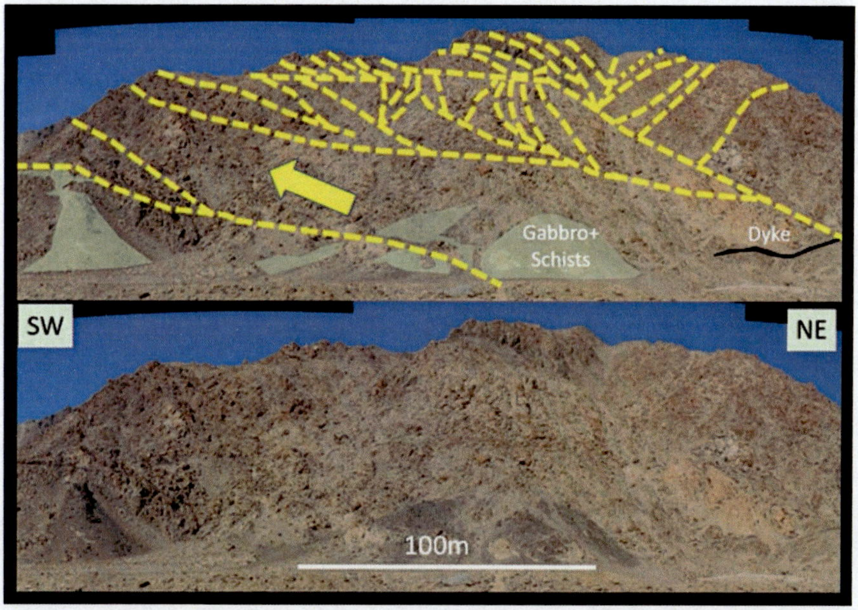

Fig. 5.8 Upper Taroo valley thrust zone composed of granite, gabbro + semi-pelitic schists, and basic dykes. The cliff face exposes a series of thrust sheets, some 20–30 m thick and exposed over a length of c. 350 m. Thrust directions are towards the left/SW, with minor back-thrusting

Fig. 5.9 Shear zone in diorite, upper-central Phyang valley. Left hand side photograph gives an indication of the scale of the shear zone, which measures c. 50 m, striking parallel to the road. The shear zone is sub-vertical and characterised by a high degree of fracture–cleavage and strong gneissic foliation. Right hand side photographs illustrate local shear zone details, the well-developed foliations and segregation of mafic and felsic minerals, and the presence of extended meta-sedimentary–hornblendite xenoliths. Note the presence of felsic or hornblendic reaction rims around some xenoliths. Car, notebook and pencil for scale (4 m, 18 cm, 15 cm respectively)

Fig. 5.10 Upper Phyang sheared and strongly foliated medium-coarse grained granodioritic gneiss. Note the mafic–felsic mineral segregation fabrics, and the highly extended xenoliths (with long: short axis ratios of up to 20:1). The gneiss is cut by a fine-grained, late-stage, biotite–feldspar–quartz granite, which intrudes along a brittle fracture at high angle to the gneissic foliation. Pencil for scale (15 cm)

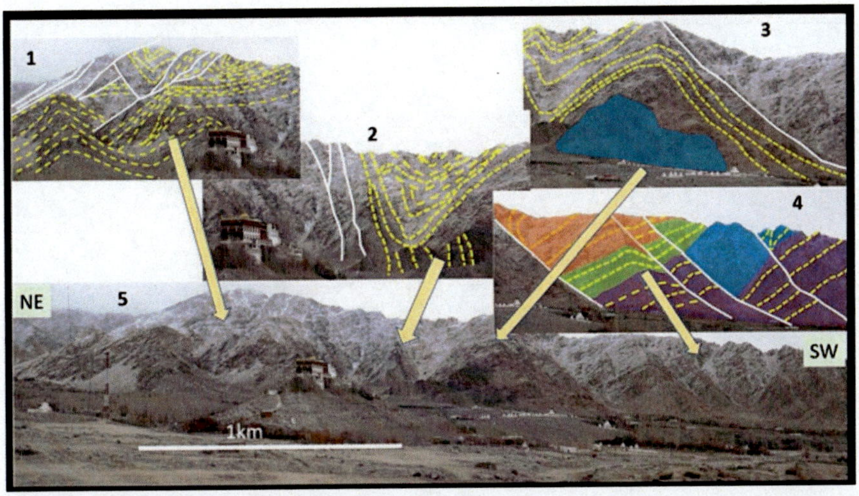

Fig. 5.11 Structure of the ridge east of Phyang Gompa. Photographic montages and interpretations (1)–(5) illustrate a range of structural features. Tight to open synforms and antiforms comprise the bulk of the c. 3–4 km length of ridge. The fold structures are cut by SW-verging thrusts to the north and SW-dipping extensional normal faults to the south, which have been reactivated as reverse faults. The antiform in Fig. 5.11 (3) is cored by a gabbro intrusion (shown in blue)

from metres to c. 150 m. The host geology varies between meta-sedimentary and granodiorite to diorite.

The upper-central Saboo valley is dominated by a complementary antiformal–synformal structure with a wavelength of several kilometres. Figure 5.13 presents a photomosaic of the northern parts of Saboo valley. Structures dip East–North-eastwards towards the NE and west–south-westwards towards the southwest in this part of the valley, forming an antiformal structure cored by granite–dioritic intrusions. Both limbs of the antiform display thrust or nappe structures with opposite vergence to one another on either fold limb. An ENE high-angle extensional normal fault cuts the antiform on the western side of the structure, and a WSW–ESE striking fault cuts the structure in the lower ground. Figure 5.15 presents a geological–deformational sequence interpreted here as: (1) deposition of sediments and intrusion of plutons; (2) ENE–WSW compression producing folding; (3) over-tightening of folds with resultant shears, nappes and thrusts verging WSW and ENE; (4) late extensional structures strike north–south and east–west.

A complementary open synform is exposed further south in central Saboo (Fig. 5.14). The synform is cut by WSW verging thrust structures.

Much of Stakmo valley is dominated by fold structures. Figure 5.15 displays a photomosaic of a synformal–antiformal structures with a wavelength of c. 900 m–1 km, and a wave height of c. 300 m. W-shaped minor folds are present in the core of the antiform on the right-hand side of the figure. These minor structures fold granite intrusions. The geology of the region comprises meta-sedimentary–diorite sequences intruded by lighter coloured granites.

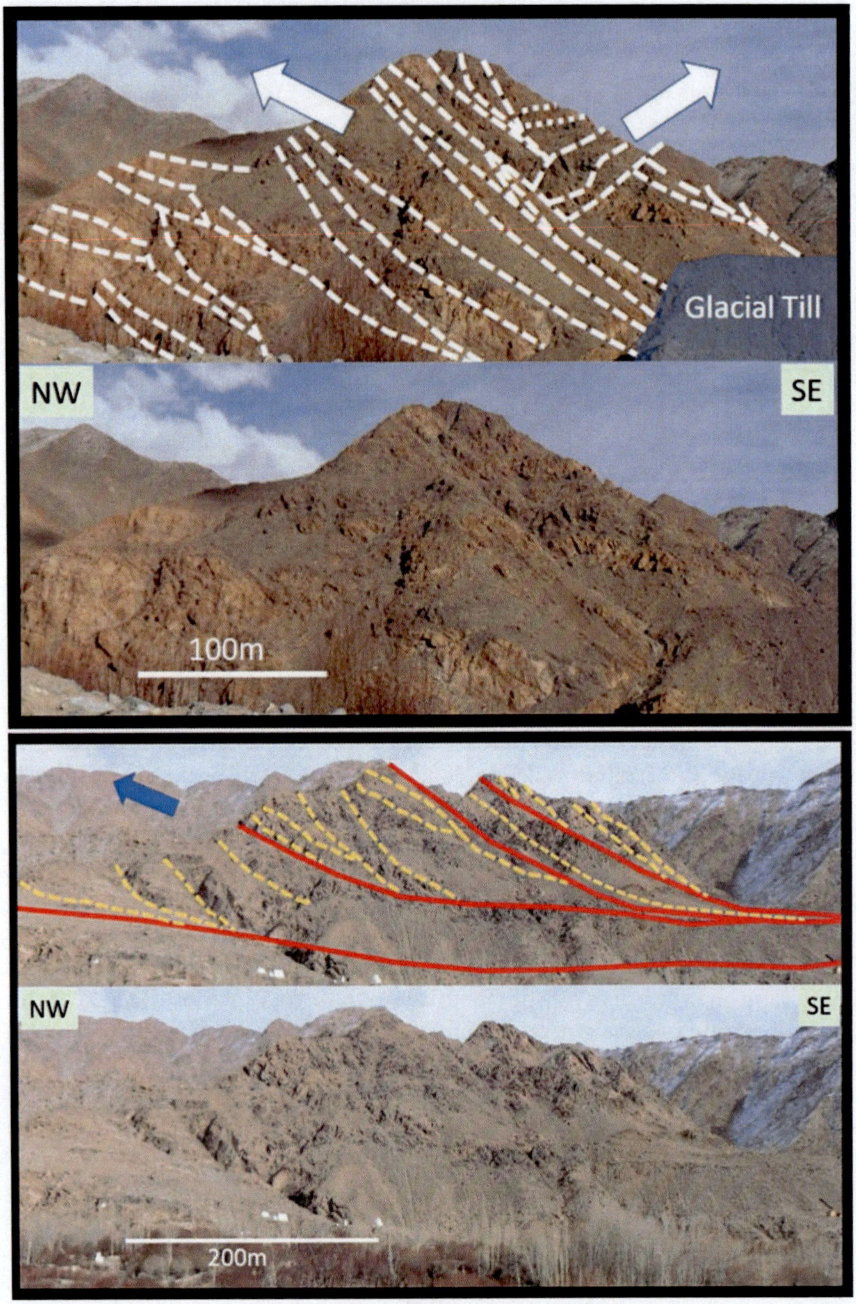

Fig. 5.12 Predominant NW-verging thrusts, with limited SE-directed back-thrusting thrust piles within meta-sedimentary and granodioritic sequence, northern Phyang

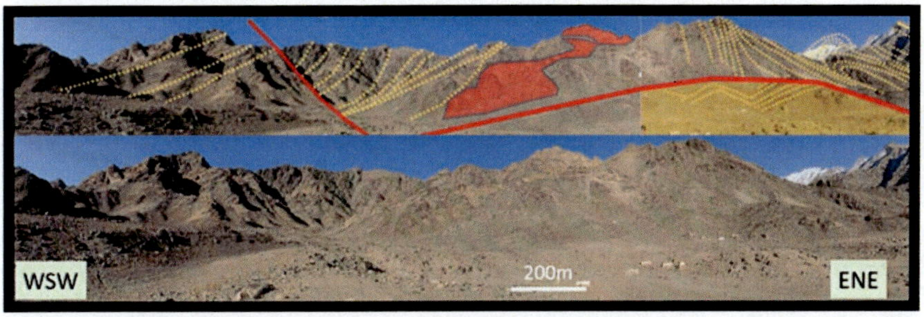

Fig. 5.13 Upper Saboo valley. Photo-mosaic displays antiformal and WSW-trending nappe structures. The antiform is faulted by a steeply dipping ENE-dipping fault on the left hand side of the figure (red), and a WSW–ESE trending fault in the central-right part of the figure. A granitoid intrusion is shown in red in the central part of the photomosaic

Fig. 5.14 Central-west Saboo valley. A broad-open synformal structure dominates the area, with sheared and thrusted diorite–granodiorite (verging WSW) present on the left-hand side of the photomosaic. Lighter coloured granite intrusions shown in red

5.2 Structural Data, Geological Cross Sections, and Structural Analysis of the N–NE Zone

The strong ENE–WSW structural grain of the High Strain Zone is less apparent within the N–NE Zone. The N–NE Zone is more complex and heterogeneous. Structural parameters are more diverse and include folds, shears, thrusts, foliations, bedding, cleavage, dyke and sill orientations and high angle faults. This contrasts with the HSZ where thrust-related structural data dominates. The only strong ENE–WSW structural grain is present within joints and topographic escarpment features, and dykes, sills and faults (Fig. 5.16) which are all expressions of brittle structures, suggesting

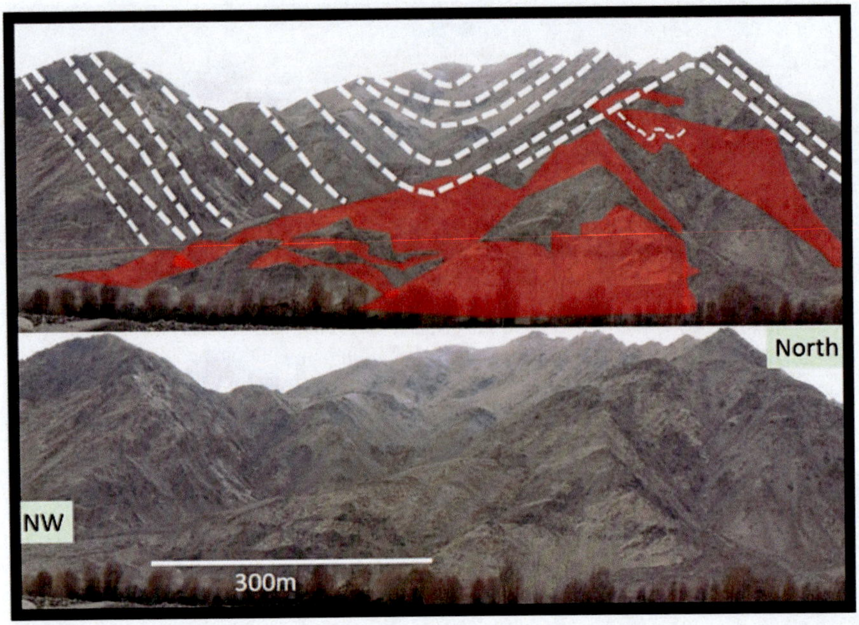

Fig. 5.15 Central-Upper Stakmo valley. Open synform–antiform structures dominate this part of the valley. Lighter coloured granitoids shown in red

that the brittle structures have a significant continuity within the whole Leh–Ladakh region. A minority of shear structures follow this ENE–WSW trend. Rose diagrams for shear zones, joints, and regional topographic features (mostly prominent and contiguous escarpments and benches), cleavage, and bedding-foliation mostly trend between NW–SE and NE–SW. Shear zones, features, joints, and cleavage planes all have strong NW–SE to NNW–SSE modes. Cleavage and contiguous joints are often parallel to or have a close angular relationship with shear zones (Fig. 5.16). Foliation and bedding data suggest fold axial plane orientations that vary between NE and NNW: larger-scale folds have axial planes-oriented N–S (Taroo), NW–SE (Phyang), WNW (Saboo) and NNW (Stakmo). If these large mountain-scale structures were combined with local field readings, the foliation-bedding rose diagram would contain a stronger NW–SE mode. This work concludes that axial plane orientations are variable across the region but are mostly oriented between NW–SE and NE–SW. Structural data differ from the HSZ with respect to dip: there is a much higher proportion of high dip structures within the N–NE Zone, particularly shear zones and related fracture–cleavage structures. Lower dips mostly related to thrust structures, similar to the HSZ.

Figures 5.17, 5.18, and 5.19 illustrate structural variations from Taroo through Phyang to Saboo and Stakmo. A strong-moderate NW–SE structural grain is clear in all regions, with Phyang and Stakmo having the most consistent structural grain with this orientation (with minor variations between E–W and NW–SE for Phyang

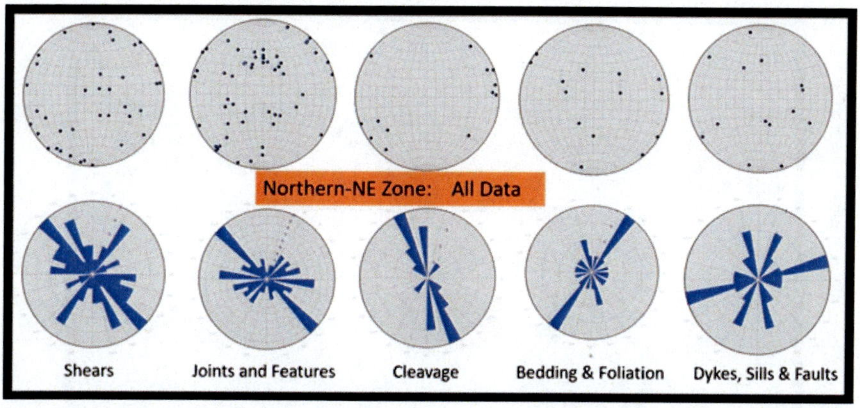

Fig. 5.16 Stereograms (poles to planes) and rose diagrams for shear zones, joints, bedding, foliations, faults, significant regional topographic features (mostly escarpments), dykes and sills for the whole N–NE zone

and N–S and NW–SE for Stakmo). Saboo and Taroo have another NE–SW mode, reflecting the presence of steeply dipping shear zones in both valleys. Taroo and Phyang contain the greatest abundance of steeply dipping structures mostly linked to shear zones and steep axial planar areas. Less steeply dipping data are mostly linked to regional topographic features and related joints/cleavage fractures, which, in turn, are strongly controlled by thrust faults. There is no systematic variation in structural grain from the NW to the SE within the N–NE zone.

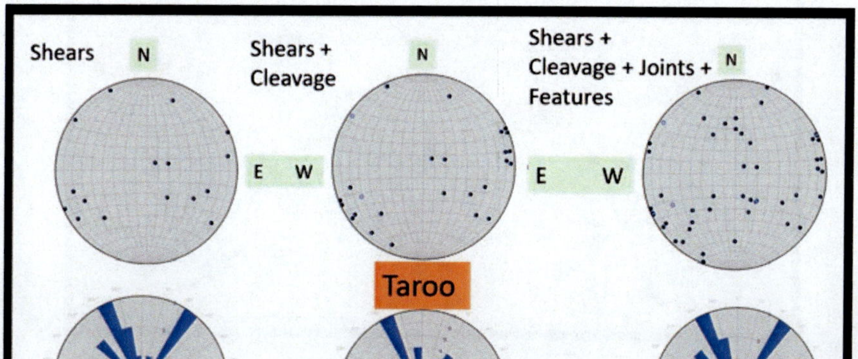

Fig. 5.17 Stereograms (poles to planes) and rose diagrams for shear zones, joints, and significant regional topographic features (mostly escarpments), for the Taroo valley N–NE zone

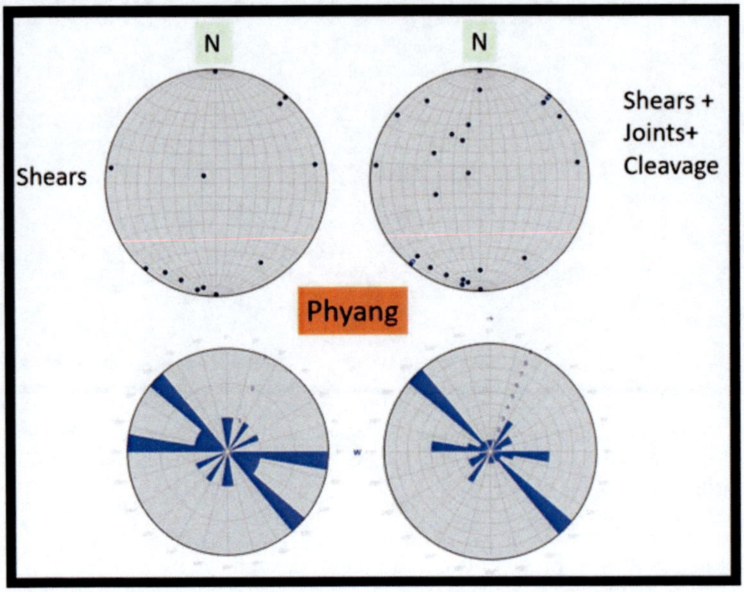

Fig. 5.18 Stereograms (poles to planes) and rose diagrams for shear zones and joints for Phyang valley, N–NE zone

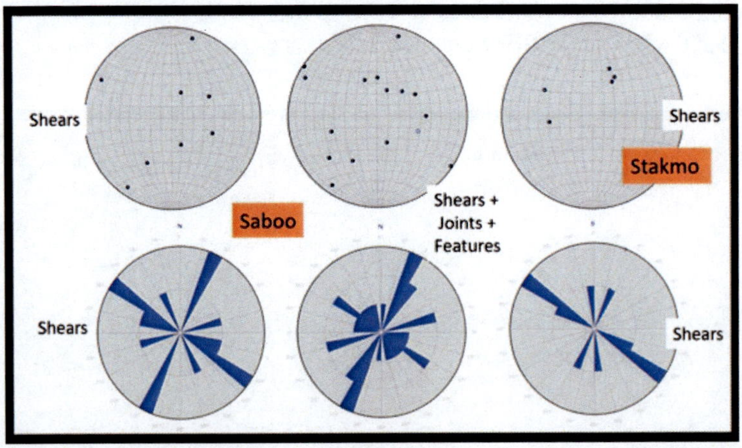

Fig. 5.19 Stereograms (poles to planes) and rose diagrams for shear zones and joints for the Saboo and Stakmo valley N–NE zone

Figures 5.20 and 5.21 present and interpret the kinematic thrust data for the whole Leh–Ladakh study region (HSZ and N–NE Zone). The addition and interpretation of thrust kinematic data in the N–NE zone must be treated with caution, as thrusting and the thrust vectors so dominant in the HSZ, lose some of their significance in

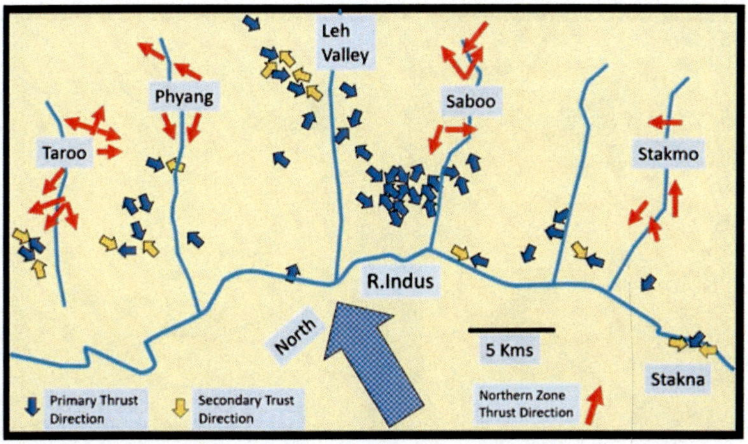

Fig. 5.20 Kinematic thrust and shear directional data for the whole Leh–Ladakh region. The N–NE Zone data added as red arrows

the more structurally diverse N–NE Zone. Nevertheless, it appears that two klippen identified from the HSZ can be extended further north in the Taroo–Phyang and Saboo valleys, with local thrust kinematic data being consistent with data within the HSZ zone. Saboo data further support the presence of the Leh–Saboo Complex Zone. Northern-most thrust directional data from Taroo, Phyang, and Saboo, indicate local reverses in thrust polarities, as well as indicating significant strike slip motions, suggesting that another structural control is coming into play in these areas. Stakmo N–NE Zone data are highly variable but do suggest that the SW vector within the Stakmo Klippen may be confined to the southernmost areas.

Schematic Geological-Structural cross sections for Taroo, Phyang, and Sabo are presented in Figs. 5.22, 5.23 and 5.24 respectively. The cross sections underscore the more diverse, heterogeneous character of the N–NE Zone with respect to the relatively simple HSZ. Steeply to moderately dipping shear zones, some km wide comprising a variable number of individual smaller width shear zones are present within the Taroo and Phyang valleys. Large to medium km-scale folding is present in all sections with fold styles varying from nappes and recumbent styles to tight-isoclinal to open are present in all sections. Asymmetrical fold vergence is towards the SW in Taroo and northernmost Saboo, and NNW in Phyang, which are in line with the klippen model analysis explained above. Thrusting and thrust structures are a component of c. 30% of the N–NE Zone, in contrast to > 90% of the HSZ. Thrust vergence is predominantly towards the SW in the Taroo and Leh–Saboo Complex Zone areas and towards the NNW in most of Phyang, with SSE directed back-thrusting in the southern part of Phyang N–NE Zone. Extensional normal faults, accommodating early-stage sedimentation and half grabens are present in Phyang and Saboo, some of which have been reactivated post-deformation either as reverse or normal faults. Steep cleavage zones are present within many fold cores. In Taroo later stage dykes and sills take advantage of shear and cleavage structures.

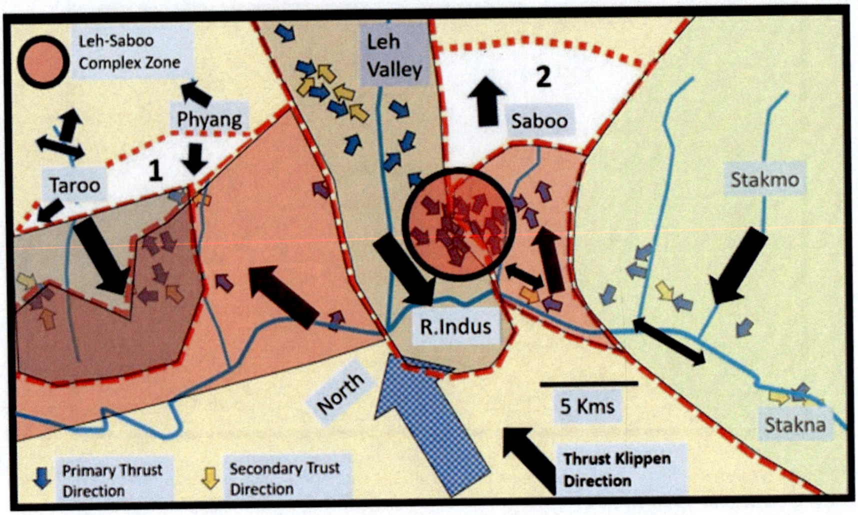

Fig. 5.21 Extended Klippen model for the whole Leh–Ladakh region. This figure considers thrust data for the N–NE zone. Extensions to the klippen identified previously (Chap. 4) are shown as Numbered areas 1 and 2. S-directed thrusts are present within the numbered zone 1 suggesting that the Phyang–Taroo S-directed klippen extends further northwards, and similarly the Saboo Klippen extends further north into the Saboo N–NE Zone. Stakmo data are less diagnostic, some Saboo data reflect the Leh–Saboo Complex Zone and northern-most Taroo and Phyang data indicate a change in thrust motions to the N–NW

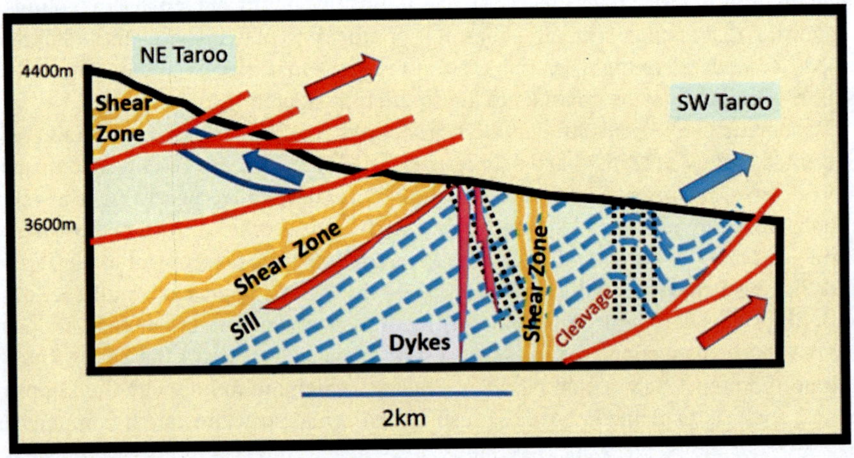

Fig. 5.22 NE–SW geological cross section through the Taroo N–NE zone

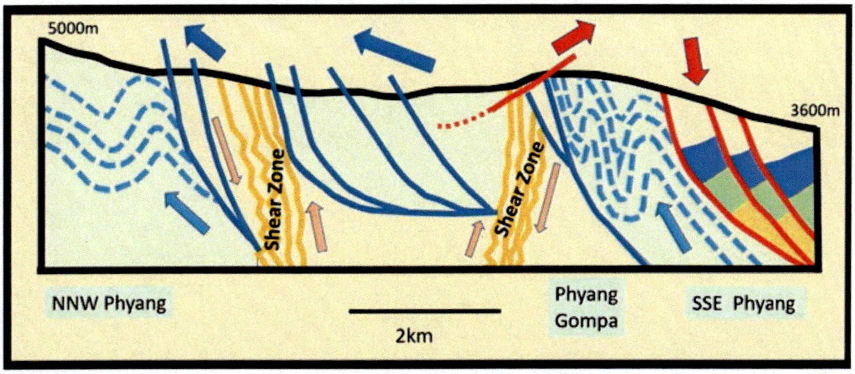

Fig. 5.23 NNW–SSE geological cross section through the Phyang N–NE zone

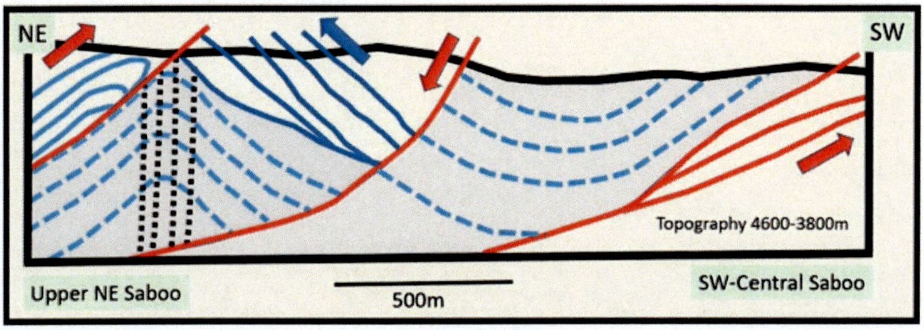

Fig. 5.24 NE–SW geological cross section through the Saboo N–NE zone

In summary the geo-tectonic temporal model for the N–NE Zone includes:

- Earliest stage sedimentation and sedimentary-graben-half graben steeply-dipping accommodation faults.
- Intrusion of the bulk of the igneous plutonic rocks.
- Main deformation event producing folds and shears at deeper crustal levels and thrusts at shallower crustal levels.
- Pene-contemporaneity of later stage shear and thrust deformation at the ductile–brittle transition zone with shears cross cutting thrusts and vice versa.
- Intrusion of later stage dykes and sills within shear, fracture, and cleavage zones.
- Late-stage reactivation of high angle extensional faults which respond as reverse or normal faults.
- Latest stage intrusion of dykes and sills associated with high angle fault reactivation.

Chapter 6
Structure and Lithology of the Indus Suture Rocks (*Indus Formation*) in the Leh–Ladakh Region

Abstract The Indus Suture Rocks crop-out south of the Kohistan batholith over a c. 40 km strike width from NW to SE. The Indus Suture Rocks are exceptionally well exposed, forming cliffs and mountains with a c. 2000 m + vertical relief in places. The unit is well bedded and contiguously bedded. Thrust and fault structures are clearly visible in mountainside outcrops. There is a wide variation in clastic lithological character from coarse conglomerates to shales. This stratigraphy constitutes a molasse sequence which, in part, onlaps unconformably onto the Ladakh batholith. Outcrops in the north-westernmost part of the study area are dominated by cyclical graded coarse sandstone/conglomerate—shale sequences akin to hemipelagic turbidites. Parallel bedded sequences of conglomerate, gritty sandstone, siltstone and shale predominate with individual beds varying in thickness between a few centimetres and 60 cm. The rocks are metamorphosed from very low grade to phyllite-low greenschist grade. Sedimentary structures include graded bedding, cross bedding, flute structures, soft sedimentary deformation structures, and erosive bases. A high abundance of mafic and feldspar minerals in some sandstones are indicative of volcanic and plutonic source rocks. Structural studies of the Indus Suture Rocks largely, from photo-mosaic interpretations, reveal complex thrust nappe/horse stacks, antiformal thrust-stacks, and related fold structures. Individual thrusts vary from 10s to 100s of m thick. Whilst there is a predominant NE-trending thrust vergence, back-thrusts verge towards the NW and SE, with local variations in thrust kinematics.

Keywords Indus Suture Rocks · Turbidites · Photo-interpretation

6.1 Scope of This Chapter

This chapter focuses on the non-batholith Indus Suture Rocks (Indus Formation) within the Leh–Ladakh area. Figure 6.1 displays the outcrop area of this sedimentary–meta–sedimentary sequence, together with the location of detailed lithological fieldwork (location 1 in the NW part of the Indus Suture Rock Unit).

M. G. Petterson, *Himalayan Thick-Skin Basement Deformation of the Ladakh Batholith, Leh-Ladakh Region, NW India*, Springer Geology,
https://doi.org/10.1007/978-3-031-31566-4_6

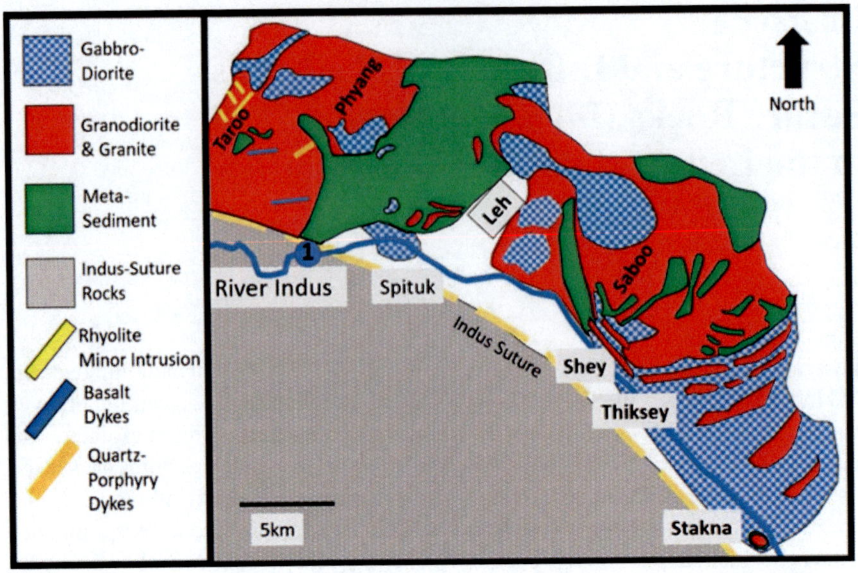

Fig. 6.1 Geological map of the Leh Ladakh region: the grey stippled zone shows the outcrop area of the Indus Suture rocks in the Leh–Ladakh area. The blue circle with a number 1 inside indicates the area of observations for lithological character

6.2 Lithological Characteristics of the Indus Suture Rocks in the Leh–Ladakh Region

Figure 6.2 presents a mountain scale view of the Indus Suture Rocks close to a significant meander in the River Indus marked in Fig. 6.1 by the blue circle. The rocks form prominent grey-green crags that stand proud within the mountain landscape and generate prodigious amounts of scree. The crags in Fig. 6.2 form bedded units dipping SSW. Figures 6.3, 6.4, 6.5, and 6.6 present examples of the Indus Suture Rocks close to this locality with Fig. 6.7 summarising key lithological features in graphic logs.

Outcrops of the Indus Suture Rocks (ISR) observed in this study are predominantly bedded sequences of sandstone, grading from conglomerate to gritty sandstone and siltstone to shale. Bedding thicknesses mostly vary between cm and dm to c. 50–60 cm. The rocks are metamorphosed between very low grade to phyllite-lower greenschist grade, and upper greenschist grade, although locally the metamorphic grade can increase to amphibolite. Rock mineralogy is dominated by quartz, feldspar, and hornblende. Some sequences are arkosic in nature, and the relatively high abundance of mafic minerals and occasional sequences of fine-grained quartz–feldspathic layers, together with a local strong green colour, is suggestive of a volcanic input into parts of the sequence.

Fig. 6.2 Part of the field area where lithological observations of the Indus Suture rocks were made. Checked white lines indicate bedding of the local units. Mountains are 3500 m above sea level (asl) and River Indus is locally at 3200 m asl

Fig. 6.3 Detail of coarse gritty sandstone, typical of the Indus Suture Rocks in the Leh–Ladakh area. Note the poorly sorted nature of the sandstone, with clast shapes varying from angular and sub-angular to sub-rounded. Quartz, feldspar, and amphibole form the main mineralogy, with quartz being dominant. This outcrop is veined with thin quartz and microgranite veinlets. Pencil is 5 cm long in the photograph

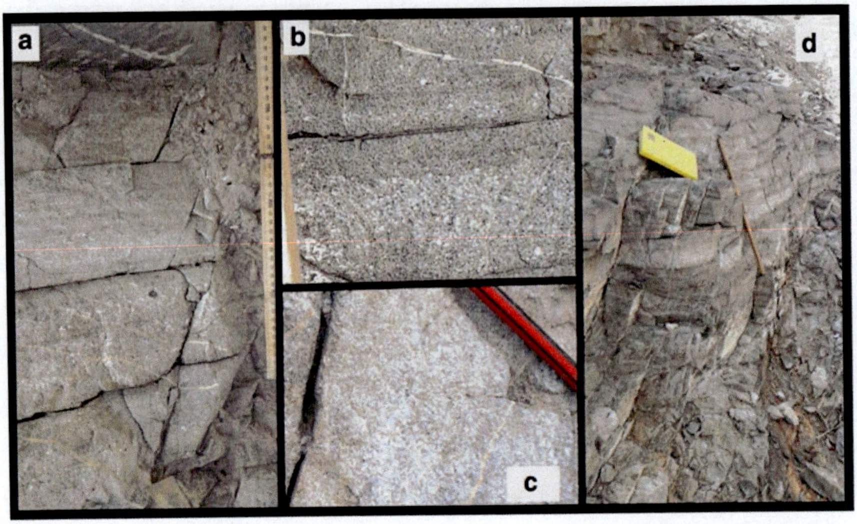

Fig. 6.4 ISR lithologies. **a** Presents rhythmically bedded coarse gritty to fine grained sandstone. Graded bedding from coarse sandstone to siltstone is apparent in some beds. Some units show erosional bases, whilst other have planar lower bases: the rocks are the right way up. **b** Presents a close-up of two units with a lower coarse gritty sandstone, grading upwards to a medium grained sandstone and shale. The upper unit is also graded and displays larger sub-rounded quartz clasts. **c** Shows details of a better-sorted coarse-medium quartz-rich sandstone. **d** Presents several sub-horizontal beds cut by steeply dipping, closely-spaced-cleavage. Various scales are presented: ruler with cm scales, a c. 4 cm length pencil, and a 17 cm notebook

Individual beds of the ISR are mostly poorly sorted, varying at least between coarse and fine sandstone, but more often between gritty sandstone or even conglomerate and siltstone to shale. There are occasional well-sorted units. Bedding is mostly parallel bedding with hints and strong expressions of cross-bedding. Bases to beds are planar to erosive with the upper beds eroding flute structures within lower beds. Way-up-criteria, such as graded bedding, cross bedding, and erosive bases, indicate that the observed units are the right-way-up.

Clasts vary in shape between angular-sub-angular and sub-rounded. No highly rounded clasts are present. Quartz is the dominant modal mineral. Rocks are clast-rich, locally clast-supported, with a silica cement between grains. Conglomerates contain larger clasts that are 8–30 mm in size, set in a gritty-sandstone matrix. Shales tend to be thin and form a minor part of the sequence. Lenses of larger clasts several cm long (e.g. grit grade) are present within some beds, isolated within a finer grained matrix.

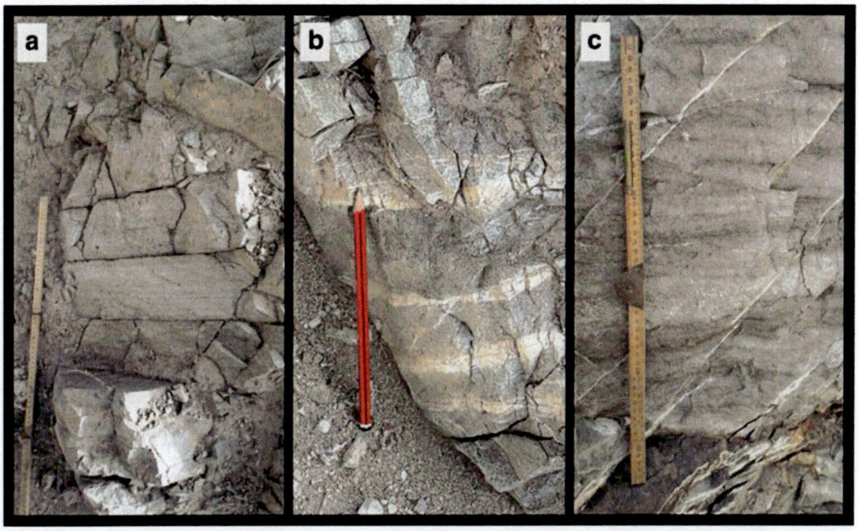

Fig. 6.5 **a** Presents 6 individual beds of Indus Suture Rocks, mostly with sharp contacts. Note the parallel and cross bedding structures present in the middle beds. **b** Shows rhythmic variations of coarse-medium grained sandstone bases grading up to fine siltstone pale coloured layers, possibly of volcanic origin. **c** Presents graded beds with coarse sandstone bases and siltstone to shale upper beds, cut by a regular series of quartz veinlets. Scales include ruler (1 m full length) and a 14 cm long pencil

Whilst many ISR outcrops retain a predominant sedimentary character, other outcrops are significantly re-crystallised, and metamorphosed to phyllitic–amphibolitic schists. Hand-lens textures are crystalline rather than clastic. Some rocks take on a strong phyllitic sheen reflecting the growth of greenschist minerals. Rocks are commonly cleaved with closely-spaced (mm–cm) cleavage within high-strain zones, and less intensely cleaved rocks with fracture–cleavage spacing of 5–10 cm are common. Less regular fractures are present alongside cleavage structures, with outcrops breaking up into hackly-fractured rock. Many outcrops are riddled with anastomosing veins and veinlets, measuring mm-10 cm wide. Other outcrops have cleavage planes occupied by veins. Silica and quartz veins are common as are micro-granitic micro-intrusions. The ISR is locally folded and faulted, with thrust faults being the most common brittle structure.

Fig. 6.6 The Indus Suture rocks within the study area are predominantly metamorphosed to phyllite and chlorite grade. **a** Presents a typical bedded, cleaved, and fractured phyllitic sandstone, riddled with anastomosing quartz and microgranitic veinlets. **b** Presents a strong phyllitic sheen within an altered and veined meta-sandstone. **c** Presents pervasive silica and microgranitic veining and intrusion with related alteration within a meta sandstone. Rulers are 25 cm, the glove in **a** is 15 cm long

6.3 Structural Interpretation of the Indus Suture Rocks (ISR) in the Leh–Ladakh Region

Figure 6.8 shows the locations of Figs. 6.9, 6.10, 6.11, 6.12, 6.13, and 6.14 which are interpreted photo-mosaics or individual photo-images taken of the northern part of the Indus Suture Rock sequence. Images are presented from a vantage point of the Ladakh batholith looking mainly south, south–west, and west, or from within the ISR outcrop area. The images are interpreted through (1) highlighting the main bedding planar structures, which, in turn, allow a (2) structural interpretation to be made of various units within the ISR based on bedding dip, strike, and overall structural architecture, together with related folding and thrust-stack structures. The ISR are a well bedded sequence. Unlike the intrusive rocks within the Ladakh batholith, this strong pervasive 3-D fabric is particularly prone to deformation and is able to present clear visual images of the styles of deformation. It is more difficult to 'see' deformational styles and geometries within plutonic rocks, which have no or few primary planar structures: deformation must first produce these in the form of

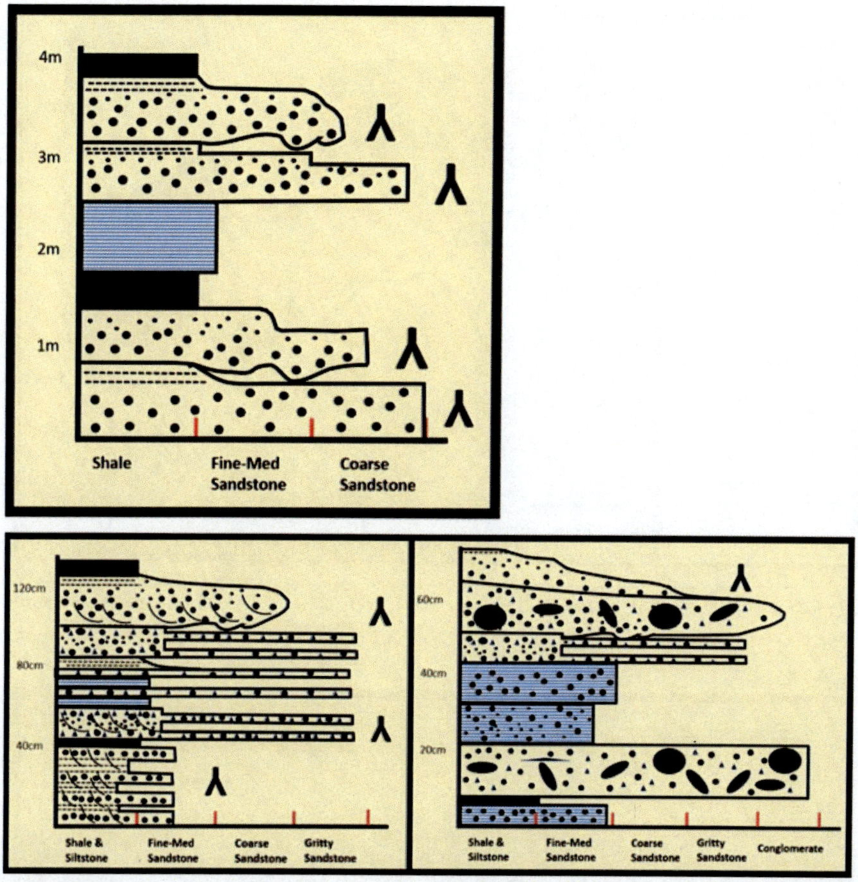

Fig. 6.7 Graphic lithological representation of measured stratigraphic logs of the Indus Suture rocks. Bedding varies between 10 and 60 cm. Graded bedding is common with coarse gritty sandstone bases and fine sandstone to siltstone-shale tops: Y symbol indicates way-up. Rocks vary from conglomerate to siltstone and shale in grain size. Cross bedding is present in within some units. Some beds have erosional bases

cleavage and metamorphic foliations before deformation structures become clearly visible.

The general picture that emerges from the ISR image analysis is one of 1000s of m of bedded strata deformed within a series of thrust units and related fold structures. Individual thrust units vary from 10s to 100s of m thick. 'Form' lines within thrust stacks identified from careful tracking of bed morphologies as the beds cut across-strike within several kilometres of mountainous terrain, reveal a range of thrust-related morphologies including: (1) discrete units behaving with structural integrity and cohesion within an individual thrust stack, significantly different to surrounding/adjacent units; (2) elements of upwards-divergent, downwards-convergent structures rooting from a parental structure; (3) cross-cutting fault and

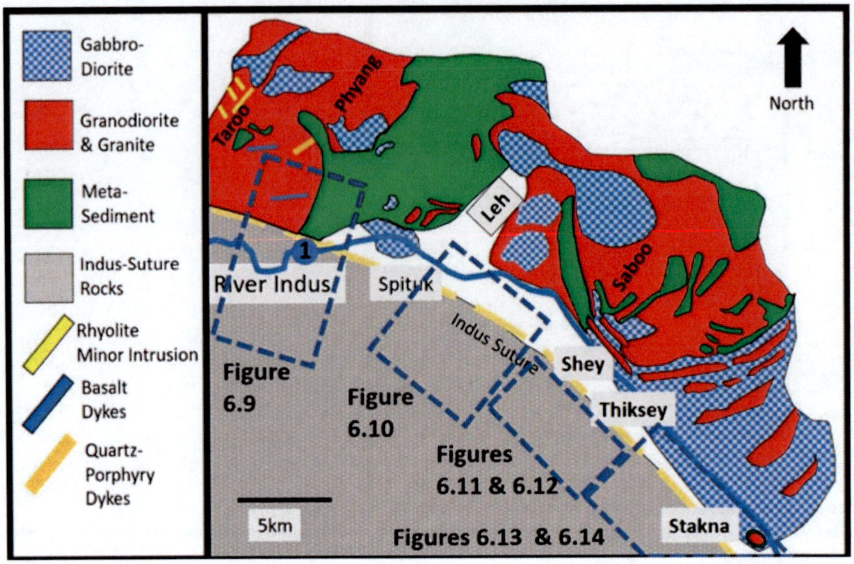

Fig. 6.8 Location of photomontages (Figs. 6.9, 6.10, 6.11, 6.12, 6.13, and 6.14) discussed in the text

Fig. 6.9 View looking south from the Ladakh Batholith in Phyang valley towards the Indus Suture rocks. Indus Suture rocks are strongly bedded and reveal several thrust units, shown here in different colours. Thrust directions are towards the N–NNE and NE. Mountains are up to 4500–5000 m above sea level (asl) with the local Indus valley at 3500 m asl. Individual thrust units (numbered 1–4) are 100–250 m thick

thrust unit relations; (4) units with different (back-)thrust directions; (5) thrust-related antiformal and synformal structures, many of which have directional kinematic vergence; (6) closely stacked horse-thrust structures with related sole and roof thrusts and clear senses of kinematic directional indicators.

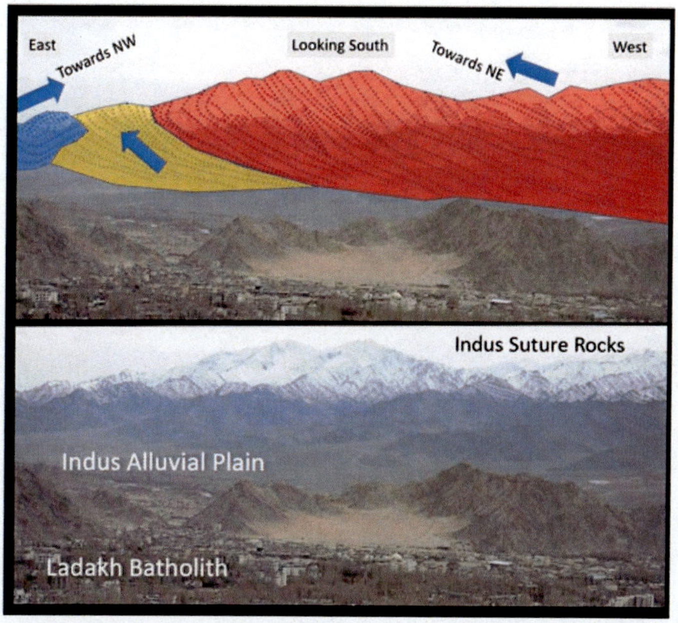

Fig. 6.10 View looking south from Tsemo Gompa, Leh valley. Mountains are up to 4500–5000 m asl and the local Indus river is at 3200 m asl. Coloured zones are interpreted as units with internal structural integrity. Two thrust units (red and yellow) are NE-directed whilst the blue unit represents a local NW backthrust

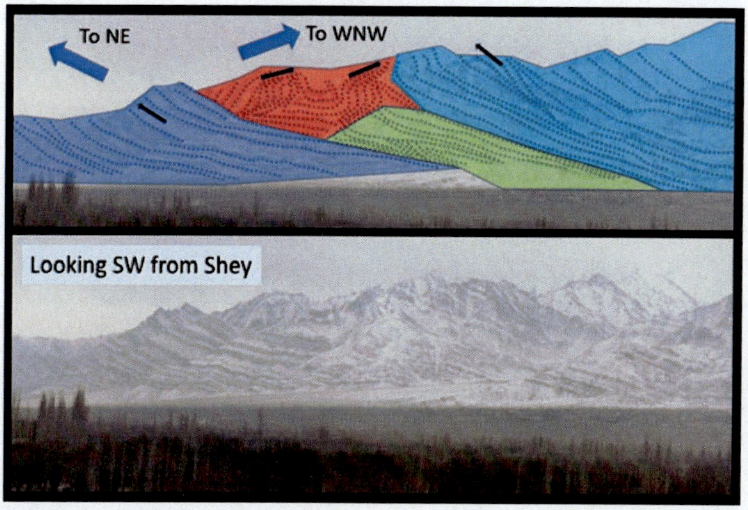

Fig. 6.11 View SW from Shey Gompa. Mountains are up to 4500–5000 m asl and the local Indus river is at 3200 m asl. The Indus Suture rocks form thrust units 100s of m thick that predominantly are NE-directed. A complex zone (in red) comprises a series of thrust-induced folds and stacked thin thrusts with bounding roof and sole thrusts: these suggest a back-thrust local motion towards the WNW

Fig. 6.12 View south from Stakmo valley with Thiksey Gompa in the foreground. Mountains rise to 4500–5000 m asl with the local Indus river at c. 3000 m asl. Thrust units 100s of m thick predominantly are NE-directed. There is a highly complex structural zone to the left of Thiksey Gompa (in blue) which presents a series of antiforms and synforms, often broken up by intervening thrusts. Steeply dipping units in this area may represent thrust stacks/horse structures, which, along with fold vergence data, suggests possible thrust movements in both the predominant NE direction and a back-thrusting NW direction

Ladakh presents opportunities to examine 100s of m of strata over kilometres of desert terrain displaying almost complete rock exposure locally. Valley floors are mostly at 3000–3500 m with the highest mountain peaks at 4500 to > 5000 m allowing the opportunity for the study of c. 2000 m of vertical exposure.

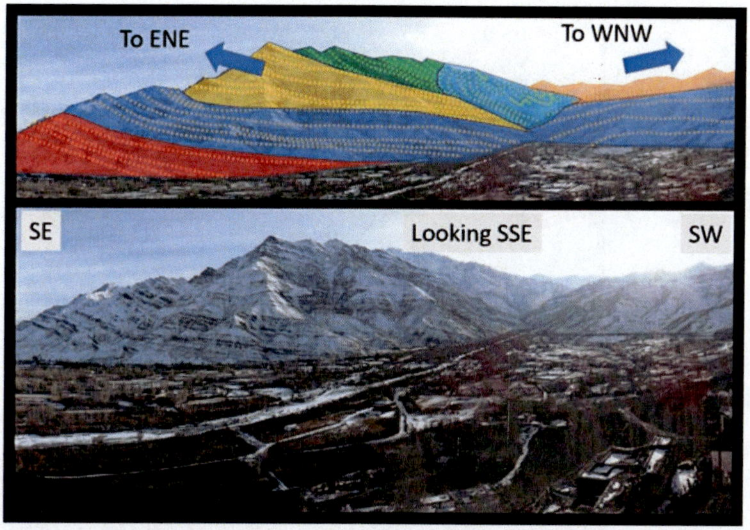

Fig. 6.13 View to the SSE from Matho Gompa, situated within the Indus Suture rocks. Mountains are c. 4500 m asl and the local valley floor is 3600 m asl. Thrust stacks, 100–200 m thick, mostly are directed towards the WNW with one unit thrusting towards the ENE

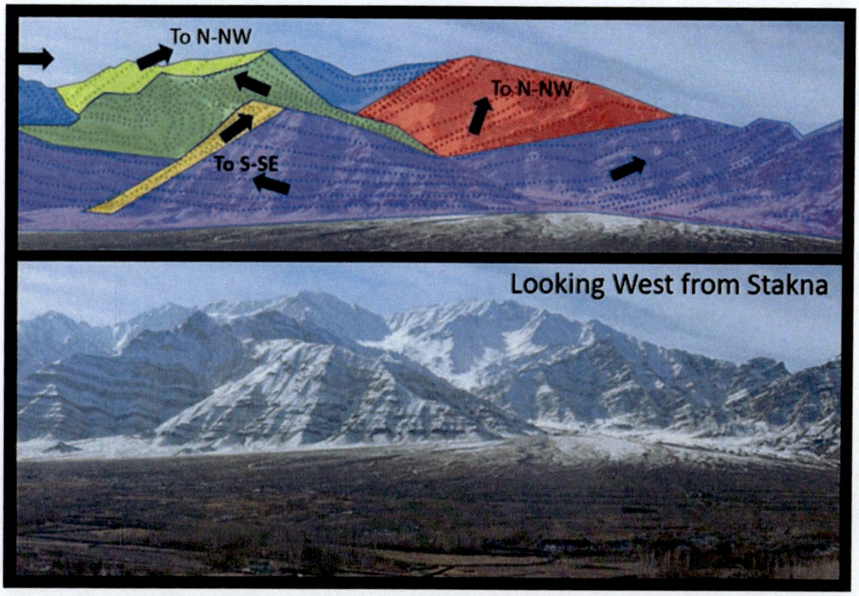

Fig. 6.14 View SW from Stakna Gompa which includes Matho valley on the right-hand side of the photo-mosaic. Mountains are 4500–5000 m asl and River Indus is locally at 3200–3300 m asl. The image interpretation presents a series of thrust stacks 10s to 100s of m thick, with varying directional trends, from SE to North to NNW

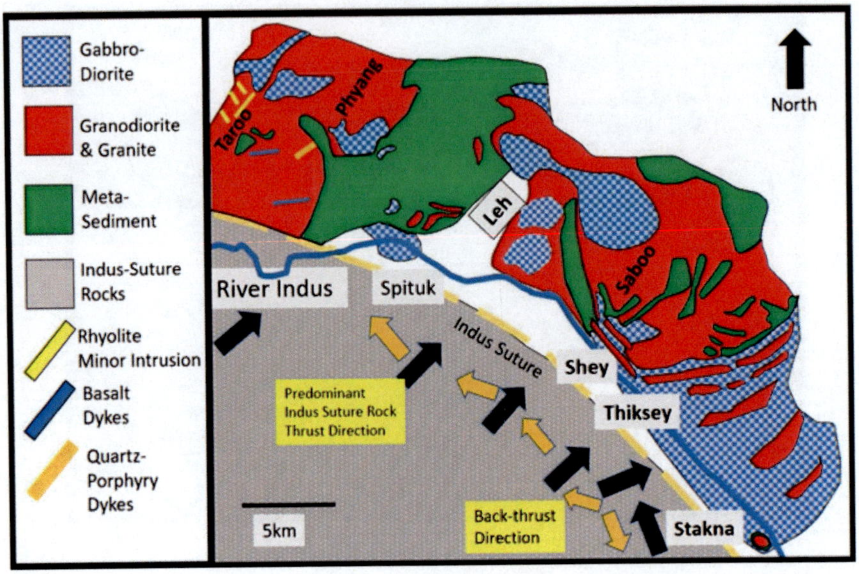

Fig. 6.15 Map of predominant thrust (black arrows) and backthrust (orange-brown arrows) directions from interpreted image analysis within the Indus Suture Rocks in the Leh–Ladakh region

Key conclusions from the study of ISR deformation reveal: (1) numerous individual thrust units, 1000s of m thick, with their own unique directional indicators; (2) common kinematic thrust and fold vergence directions indicating predominant movements towards the North–Northeast, and locally west–north–west; (3) elements of contra-back-thrusts with movement directions mostly towards the N–NW and occasional ENE; (4) complex geometries including tight-recumbent folds, dissected folds, and repetitive thrust stacks which mostly coincide with thrust convergence zones.

Figure 6.15 summarises, in map-format, the key results from the structural analysis of photo-mosaics of the Indus Suture Rocks within the Leh–Ladakh region. In most of the region the predominant thrust direction is towards the North–East, approximately 90° from the regional strike direction. Back-thrusts are mainly towards the NW or WNW. Between Thiksey and Stakna thrust directions locally change within predominant thrusting towards the NW and ENE and back-thrusting towards the NW and S–SE.

Chapter 7
Deformation Model for the Ladakh Batholith in the Leh–Ladakh Region

Abstract Magmatic activity within the Leh region is dated at 70–45 Ma, although this study predicts that this range does not reflect the full temporal spectrum of magmatism. Crustal depth pluton emplacement estimates range between 7 and 15 km. Low temperature geochronological studies indicate the batholith was uplifted to c. 3 km between 8 and 30 Ma. Zanskar structural studies indicate active cover-basement thrusting between 45 and 25 Ma with cover deformation in the Leh area to at least the Pliocene. Structural studies of the Nanga Parbat Syntaxis indicate Recent-present day active tectonics and the presence of the brittle–ductile transition zone at depths of c. 5 km, with an equivalent brittle–ductile transition zone depth of 12–6 km in the Leh area. Results from this study suggest a fractured basement crustal block deformation model with individual blocks exhibiting unique deformational histories, significant vertical and lateral movement, and variable depth exposure. Basement plutonic deformation contrasts with cover deformation in part due to the lack of pre-existing contiguous planar structures, the variable nature of stress transmission, and wide range and variable character of pre-existing fractures, weaknesses, and anisotropies. A tectonic-deformation model is presented exhibiting: (1) intrusive activity (Cretaceous–Eocene); (2) extensional rift deformation penetrating basement (Eocene–Oligocene); thrust-shear deformation within a fractured basement (Eocene–Miocene) and; (4) North-directed thrusting affecting cover and basement (Miocene–Pliocene).

Keywords Crustal blocks · Fractured basement · Brittle–ductile transition zone

7.1 Geotectonics Structure and Deformation

Published literature presenting structural, deformational, intrusive age data, low temperature chronological data, and sedimentology for the Indus Suture Rocks provide a range of constraints for a deformational model of the Ladakh batholith (Table 7.1). These literature data complement and help interpret primary data presented in the current study.

M. G. Petterson, *Himalayan Thick-Skin Basement Deformation of the Ladakh Batholith, Leh-Ladakh Region, NW India*, Springer Geology,
https://doi.org/10.1007/978-3-031-31566-4_7

Table 7.1 Constraints from the literature/this study and implications all used to develop a deformational model for the Ladakh batholith in the Leh region

Literature constraints	Study constraints/implications
Ladakh batholith intrusion ages, Leh area, 70–45 Ma (Cretaceous–Eocene), mostly 62–45 Ma (Palaeocene–Eocene)	Batholith mapped as a heterogeneous intrusive unit with large plutons, dykes, sills, veins and compositions from mela-gabbro to leuco-granite. Most areas display a chronology of events
Estimated depth of intrusion for Ladakh–Kohistan batholith plutons, 7–35 km, with 7–15 km most likely for Leh and northern Kohistan batholith areas	Pre-intrusive sedimentary units present locally. Cyclical sedimentary units with graded bedding, likely turbidites. Now mostly schistose to locally gneissose
Ladakh batholith clasts present within Indus Suture Rocks aged 50–24 Ma (Palaeocene–Miocene). Unconformity on the Ladakh batholith Palaeocene–Upper Eocene, Oligocene, and Early-Miocene (diachronous)	Parts of the batholith surfaced, and were eroded from at least 50 Ma (Palaeocene), although this varied from place to place (Eocene–Miocene). Other parts (now exposed) remained deep within the crust
Hemis Formation (Indus Suture Rocks) contains boulders up to 3 m, and channels cutting 10 m into Ladakh batholith in the Leh–Hemis area (Eocene). Eocene–Oligocene period particularly rich in Ladakh batholith clasts	Rapid uplift/denudation of the batholith, Eocene–Oligocene times (50–35 Ma)
Indus Suture Rocks deposited within a 20 km wide intra-montane basin. Deepest sediment burial 3–5 km. Basin had an active life of c. 50–6 Ma	Extensional structures forming intra montane rift may have penetrated the basement and/were assisted by basement extensional structures identified in this study. Proto-Indus River formed from c. 30 Ma
Uplift through 150 °C isotherm (depth of c. 3 km with a 40 °C/km geothermal gradient) at 30 Ma (Oligocene), 20 Ma (Miocene) and 8 Ma (Pliocene) for southern, central, and northern batholith sections respectively	Much of the study area reached c. 3 km and 150 °C between 30 and 20 Ma. This could be at least part—facilitated by basement thrusting, and variable movement between basement blocks. Basement block model suggests that local uplift/denudation may have been diachronous
High denudation rate 15–5 Ma in the north, southern batholith buried by Indus Confluence sediments	Burial of eroded batholith by sediments and Indus Molasse would have impacted geothermal gradients, isotherms and denudation processes
Thrusting from Late Cretaceous–Pliocene (70–5 Ma). Zanskar: SW directed thrusting 45–25 Ma (Eocene–Oligocene) with north-directed thrusting from 15 Ma (Miocene–Pliocene). Foreland and intramontane basins from 50 Ma	South and north-directed thrusts present in the batholith. Some south-directed thrusts may be part of the earlier cover-south directed thrusting, but others are probably too young as they cut the predominant north-directed thrusts, interpreted at least in part as contemporaneous with the 15–5 Ma N-thrusting of the Indus Suture Molasse

(continued)

Table 7.1 (continued)

Literature constraints	Study constraints/implications
Shyok Suture/Thrust mostly south-directed, Main Karakoram Thrust right-lateral transpressive	Whilst possibly too distant to be a strong control in the Leh area, south-directed thrusting and a constraining buffer-presence may influence basement structural process
Nanga Parbat syntaxis deformation includes basement ductile and brittle deformation, pop-up, divergent, and vertical stretching gravity-channel structures. Brittle–ductile transition zone at 5–6 km (c. 100–120 °C/km geothermal gradient)	Batholith textures vary from granular/undeformed through to fractured/cleaved and brittle-deformed, to strong penetrative shear fabrics and ductile-deformation related. Thrusts, pop-ups, flower structures, cataclastic rocks, extensional zones, listric faults present in the batholith basement
Plastic deformation studies for quartz in the Ladakh batholith indicate required temperatures of 300–500 °C	Deformation styles presented in this study require this temperature range for process activation and maturation (thrusting/shearing)

Literature considered in detail in Sect. 7.2 below. Key findings presented here for discussion and model development purposes. See text in this Chapter, and Chapters 1, 2, and 3 for reference details.

Key constraints for a new deformational model in the Leh–Ladakh region include (Table 7.1): (1) a local span of intrusive ages from 70 to 45 Ma, mostly 62–45 Ma; (2) estimated depths of intrusion from the Ladakh and north Kohistan areas of around 7–15 km; (3) a variety of data from the Indus Suture rocks suggesting the presence of an active intra-montane basin. The basin received sediment from c. 50–6 Ma, accommodating a sediment pile at least 5 km thick, accessing abundant Ladakh batholith clasts, particularly during Eocene–Oligocene times, with a diachronous unconformity with the Ladakh batholith dated at Palaeocene–Upper Miocene; (4) low temperature chronological age data suggesting that the batholith reached depths of c. 3 km (through uplift/denudation) between 30 and 8 Ma from south to north; (5) structural studies suggest the presence of active thrusts in the south Zanskar region from 45 to 25 Ma, with upright structures in the lower Zanskar valley, and north-directed backthrusts involving the Indus Suture molasse from c. 15 Ma; and, (6) research within the actively tectonising Nanga Parbat syntaxis demonstrates how crystalline basement displays a wide range of structures from thrusting and folding, to pop-up structures, cataclasites, and a range of brittle–ductile fault relationships between the basement Indian Plate and Kohistan–Ladakh terrane rocks. The Nanga Parbat area brittle–ductile transition zone is estimated as shallow as at c. 5 km depth, with a high local geothermal gradient of 100–120 °C per km (by extension c. 12–16 km depth in the Leh region with a c. 40 °C per km geothermal gradient, as estimated for the Leh area by Kirstein et al. 2009).

Figure 7.1 presents the broader plate-level structure for Ladakh. The KLA is bound to the north by the Shyok Suture, and to the south by the Indus Suture. The Shyok Suture separates the KLA from the Karakoram terrane and is interpreted as a south-directed steep shear-fault, also associated with the right-lateral transpressional

Main Karakoram Fault. The Indus Suture in the Ladakh area is interpreted as a reactivated north-directed thrust system. The Zanskar regional structure changes from upright to north-verging structures in the north and south-directed thrusts in the south (with the Zanskar Shear being a several km-throw shear system separating high- and low-grade metamorphic rocks). The High Himalaya and Indian Plate are characterised by south-directed piggy-back thrust piles. The Ladakh batholith is broadly sandwiched between different structural blocks: the Karakoram terrane to the north with a south-directed thrust-shear/right-lateral transpressional contact, and, to the south, by the north-directed thrusting Indus Suture Zone, beyond which are upright, then southwards-directed thrust systems (Fig. 7.1). In Kohistan Coward et al. (1982a, b, 1986) modelled a similar structural architecture broadly as a crustal pop up with upright central structures and divergent north-directed and south-directed structures to the north and the south. The Ladakh batholith may be affected by north and south-directed thrust and shear systems, and by the northern and southern sutures. In the Leh region, the active tectonics of the Indus Suture–Zanskar zone are the likely predominant control, although at deeper basement levels there may also be an influence from a reactivated Shyok Suture/Main Karakoram Thrust (Fig. 7.1).

There are few studies of deformation of the Ladakh batholith. Most workers have researched the batholith for geochronological/geochemical/magmatic processes purposes e.g. Upadhay et al. (2008), Honegger et al. (1982), Jadoon et al. (2019), Scharer et al. (1984), Ravikant et al. (2009), Kumar et al. (2012), Kumar (2020), and Ahmad et al. (1998). Weinberg and Dunlap (2000) do discuss elements of batholith deformation in the Leh area and further north. They recognise migmatites

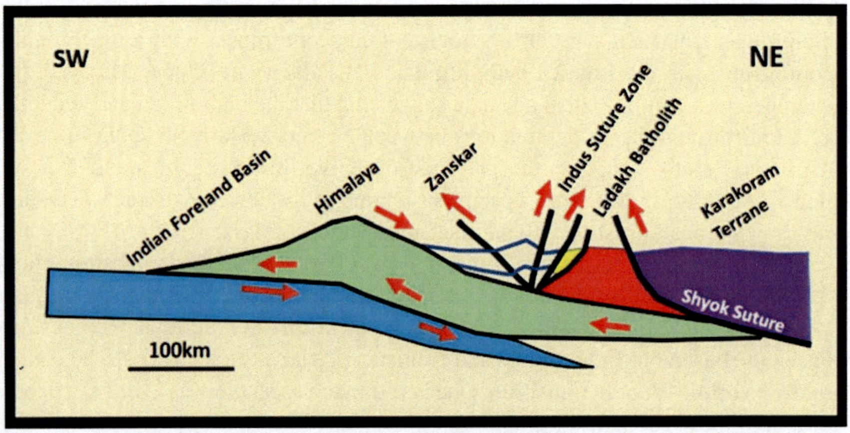

Fig. 7.1 Plate tectonic scale structural setting of the Ladakh batholith (Adapted after Coward et al. 1986; Searle and Hacker 2019, acknowledgement Geological Society of London). The Ladakh terrane is bounded to the north by the Shyok Suture and Main Karakoram Thrust, which separates Ladakh from the Karakoram terrane, and to the south by the Indus Suture, which separates Ladakh from the Indian Plate. The northern part of the Indian Plate underthrusts Ladakh/Karakoram/Tibet and is itself thrust-faulted by a series of south-directed thrusts which are a main process in forming the Himalayan High Mountain belt

and deformed plutonics around Leh and beyond, including gneisses, mylonites and ultra-mylonites. Studied microstructures and quartz plastic deformation within the batholith are interpreted as occurring at temperatures of 300–450 °C. They develop a model whereby the batholith was affected by dextral shearing, perhaps linked to the Main Karakoram fault, or the Shyok Suture, and that deformation occurred close to the age of intrusion, as rapid cooling would prevent plastic deformation of quartz to occur. Weinberg and Dunlap (2000) did not have the benefit of more recent low temperature chronological data, nor the results of deformation studies from this work, that reveal strong and widespread deformation within the batholith.

Figure 7.2 summarises the deformation styles, architecture and patterns observed and recorded in this study. The predominant deformational style is thrusting accompanied by localised shearing within parts of the thrust systems (Boyer and Elliot 1982). Thrusts moved towards the north and the south, with north-directed thrusting predominating. Significant lateral strike-slip movement accompanies thrust systems. There are a range of thrust styles, the most common of which is a classic dendritic architecture with a primary basal sole thrust from which root many secondary splay faults, that form regular, parabolic, concave-upwards geometries, with smaller tertiary splay detachments rooting from the secondary faults. This geometry has a consistent vergence allowing for measurement of thrust movements in 3-D space. Floor/basal and roof thrusts are common, often bounding piggy-back style aggrading individual thrust units/horses forming duplexes. Locally these stacked horses and duplexes may form antiformal stacks. Nappes are locally present cored by basement sedimentary and plutonic rocks. Two styles of thrust 'collision' are present within the area. In the southernmost part of the batholith divergent south-directed and north-directed thrusts produce flower-pop-up structures whilst further north opposing thrust directions form vertical pop-up and tri-shear structures. There is a complex chronology of thrusting, with later thrusts deforming earlier ones, sometimes overturning early thrust-dips, and thrust stacks at different crustal/structural levels cutting pre-existing thrusted crust. Listric faults are present, some of which are pre-existing sedimentary basinal structures that moved to accommodate pre-batholith turbiditic, probable marine sediments. These faults have been re-activated in both normal and reverse senses, probably during the last deformation period. More diffuse ductile deformation dominates certain areas of the study region. Deformation is concentrated within discrete shear zones metres-10s of m wide with several shear zones present across kilometres of exposed crust. Shear zones strongly deform plutonic rocks, developing mature gneissose and schistose foliations, and S–C style fabrics. Other parts of the study region are less deformed and comprise folded pre-batholith sedimentary rock and batholith with a high density of later granitic intrusions and minor intrusion (sills, veins, and dykes). Fold styles vary from upright and open, to closed and recumbent, with a directional vergence. The less deformed parts of the batholith are affected by normal faulting, and re-activated normal faults, with a reverse sense of movement. These faults cut early fold structures and are part the latter stages of deformation.

The more intensely deformed parts of the batholith impact upon the full range of geology present from pre-batholithic sedimentary rocks to the youngest granitic minor intrusions. The young minor intrusions are less deformed, with cleaved

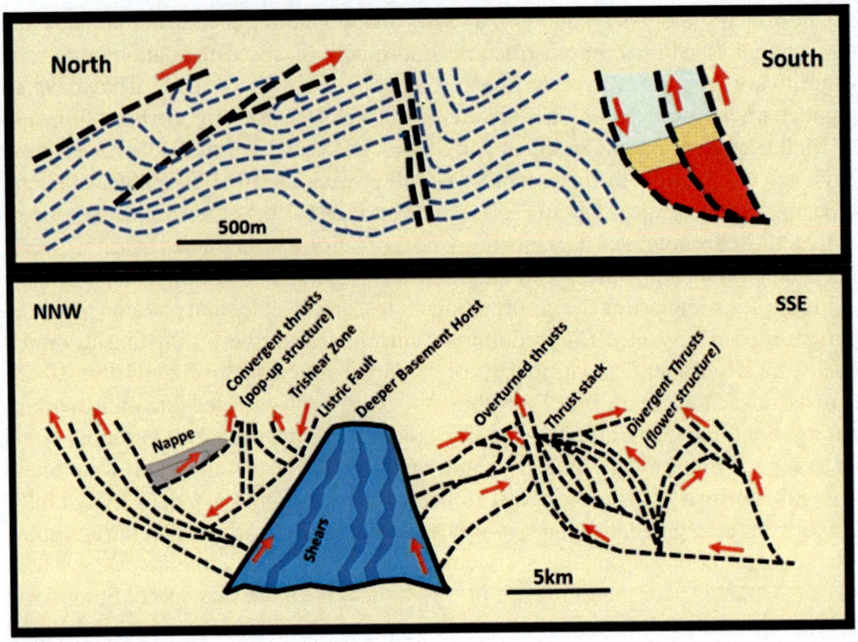

Fig. 7.2 Structural model of the basement geology of the Leh–Ladakh region. The upper diagram is typical of the shallowest levels of the basement. Pre-batholithic turbiditic sedimentary rocks are folded symmetrically and asymmetrically, with fault and thrust systems breaking up the fold structures. Asymptotic concave upward listric fault systems with unfolded stratigraphy are reactivated. The lower diagram synthesises measurements and interpretations from the Taroo, Phyang, Leh valley, Saboo, Stakna, and Stakmo sub-regions. The southernmost margin of the Ladakh batholith has a divergent thrust system with a resultant flower structure-pop-up structure. Other parts of the batholith are affected by convergent thrusts which produce 'folded' compressional pop-ups. Tri-shear blocks within the basement allow for movement vertically and laterally. Local piggy-back style thrusts stacks are present. Locally, listric faults and nappes are exposed. Thrusting directions predominate, overall, to the NNW but there are areas of strong-SSW directed thrusts. Basement highs within relatively upthrown basement blocks exhibit more diffuse ductile shear-dominated deformation

dyke or sill margins and relatively undeformed centres. They cut most of the early thrust related deformation fabrics. The level of metamorphism within the deformed batholith is mostly low to greenschist, with epidote, chlorite, and tourmaline present as veins within thrust shear and fault systems, and as replacements for feldspar and mafic minerals within the plutonic groundmass. Locally the metamorphic grade is higher and spans Upper Greenschist to Amphibolite grade with oriented metamorphic hornblende crystals growing within plutonic-sedimentary gneissose/schistose fabrics. Migmatites, mylonites, cataclasites, and in-situ partial melts are present in the most intensely deformed parts of the batholith.

Larger scale structural and tectonic constraints to inform the understanding of Ladakh batholith deformation include the timing of the Shyok and Indus Sutures at 90–75 Ma, and c. 50 Ma respectively, with likely reactivation of the Shyok Suture and related thrusts (e.g. the SW-directed Waris Thrust), and shears south of the suture (e.g. the NE-directed Khalsar Thrust) and the Main Karakoram Fault from 18 Ma (Searle et al. 1998, 1999; Rolland et al. 2000, 2009; Borneman et al. 2015). To the south the local cover sequences are deformed by south-directed thrusting from south Ladakh, and into the High Himalaya and southern Ladakh, with the Indus Molasse upright structures and north-directed thrusts from 15 Ma (e.g. Searle et al. 1988). In Kohistan the Indus Suture is strongly south-directed and the Shyok-Northern Suture upright to south and northern-directed, and deformation within Kohistan is mostly south-directed in the south, upright in the centre and north-directed towards the north, with evidence of deep crust to mantle strain in parts of Kohistan (e.g. Coward et al. 1982a, b; Burg et al. 2005).

Age constraints for the deformation include the 70–45 Ma intrusion ages, the Palaeocene–Miocene diachronous unconformity with the Indus Molasse, the existence of an intra-montane rift basin from c. 50 Ma (Palaeocene), and low temperature chronology ages suggesting the batholith reached 150 °C (3–4 km) from 30 to 8 Ma (Table 7.2 and op.cit.). Crustal depth environmental constraints include estimates of depth of intrusion in the Leh region, and north Kohistan of 7–15 km, estimates of 300–500 °C for the plastic deformation of quartz–feldspar rocks, depth estimates of 5–12/15 km for the location of the brittle–ductile transition zone with geothermal gradients of 120 °C/km and 40 °C/km respectively, and the style of plastic deformation, shearing, and related fabrics, together with the mostly greenschist to local amphibolite grade of metamorphism present within the more deformed parts of the batholith.

Table 7.2 Time event sedimentation-magmatic-deformation model for the Ladakh batholith

Time	Event
Jurassic–Cretaceous	Deposition of pre-batholith turbiditic sediments in marine basins
Cretaceous–Eocene	Intrusion of batholith at depths of 7–15 km
Eocene–Miocene	Diachronous unconformity with Indus molasse. Molasse rift basin both partly controlled by basement extensional fault and creates increased faulting and fragmentation within basement
Eocene–Miocene	South-directed thrusting with general shearing and ductile deformation at lower levels. Basement broken into discrete blocks separated by NE–SW and NW–SE faults. Highest deformation produces mylonites, cataclasites and partial melting. Pre-batholith turbidites folded, locally downthrown fault blocks with horizontal bedding preserved
Miocene–Pliocene	North–East directed thrusting of molasse. Termination of molasse deposition (Upper Miocene). Complementary north-directed thrusting in the fractured basement. Some basement blocks move relatively north accentuating fracturing
Oligocene–Miocene	Intrusion of latest acid sheets, veins and bosses intruded syn-post late cleavage deformation. Dates of youngest acid sheets predicted at < 45 Ma
Miocene–Pliocene	Local high angle faults reactivated and newly generated. Extension and compressional motions

7.2 Crustal Deformation Model: Fractured Basement

The crustal deformation model proposed here has the following elements: (1) post-intrusion timing of the beginning of deformation (post-syn 50–45 Ma or Eocene); (2) a likely crustal depth of the brittle–ductile transition zone (c. 12–5 km) for the bulk of deformation, slightly deeper (c. 17 km) within the upper ductile zone for the most intense deformation, and shallower (9–10 km), within the brittle zone for some of the thrust-dominated deformation; ambient temperatures of 350–550 °C for most of the deformation, higher (at least locally) for migmatisation/partial melting. Geothermal gradients higher than 40 °C/km would decrease crustal depth estimates; (3) tectonic scale vectors that resulted in both south and north-directed thrusting, with south directed thrusting occurring mostly in earlier times, and north-directed thrusting mostly in later times, with a degree of pene-contemporaneity, and local back-thrusting present throughout the deformation; (4) higher angle normal and reverse thrusting occurred a) pre-deformation and b) during latest deformation times; (5) uplift/denudation to 3–4 km from 30 Ma, and (6) thick skinned basement style deformation within a fractured crust.

The timing of deformation is proposed to be (1) Eocene–Oligocene for the bulk of the south-directed thrusting (2) Miocene–Pliocene for the bulk of the north directed thrusting and (3) Pliocene for high angle fault formation and reactivation of older high-angle faults. Ductile shear deformation is constrained as post-intrusion age (70–45 Ma) and may span a significant time-period. Dual-polarity thrusting requires tectonic mechanisms that move the crust south and north at different periods and probably involve resistant pinned crustal units/blocks that buffer thrust vectors and encourage back-thrusts to occur.

In Kohistan this style of basement deformation is proposed to have been produced by the sandwiching of Kohistan between the Karakoram and Indian Plates, with deformation occurring mostly from Cretaceous to Eocene times. The differences between thin and thick skinned, cover and basement, deformation also requires a degree of consideration. Structural models to the south of Ladakh into the High Himalaya are derived from thin-skin cover deformation. This style of deformation, although complex, is perhaps less challenging to model and define than basement deformation. This is due to several reasons, including the legacy of thrust deformation researchers particularly from the time of Boyer and Elliot (1982) and related workers, who defined many of the rules, nomenclature, geometries and theory of thin-skinned thrust deformation alongside classic studies of thrust zones such as the Moine Thrust Zone in NW Scotland (e.g. Coward 1980, 1982, 1985; Holdsworth et al. 2006). These mature thrust theories have been applied and refined across the world with significant success within numerous cover sequences of orogenic belts. Cover sequences mostly involve bedded sedimentary rocks which can be correlated stratigraphically. Bedding and subsequent schistose foliations are relatively easily mapped and recorded during fieldwork: individual marker beds can often be tracked across-country. Changes in stratigraphic thickness and fold and thrust geometries are more easily defined. Deformation occurs within the uppermost parts of the crust, although some structures can penetrate to the base of the crust itself. Foreland thrust-directed piggyback sequences are common although the full gambit of thrust geometries (e.g. back-thrusts, out of sequence thrusts, antiformal stacks etc.,) is present in parts of numerous orogenic belts. However, relatively simple, coherent structural geometries can be clearly defined and presented (e.g. Searle et al. 1988).

In contrast basement crustal rocks mostly comprise crystalline plutonic and metamorphic rocks with no/few well developed primary foliations alongside deeply buried sedimentary-volcanic rocks. These types of geology are heterogeneous, and do not lend themselves easily to defining clear deformation geometries over large areas, nor can they easily be correlated apart from spot-radiometric dating, or if some units have an unusually diagnostic lithological/geochemical/geophysical signature. In many orogenic belts the true basement is only occasionally exposed, and basement structure is defined more from seismic reflection and refraction studies. Often there is a decollement or detachment between cover and basement sequences and the two types of crustal unit may respond to deformation in quite different ways. Clearly defining precise deformation geometries and causal processes within deeper crustal units is challenging in most orogenic belts. The work of Coward et al. (1982a, b, 1986), Treloar et al. (1990, 1996, 2003, 2007, 1989), Treloar and Izatt (1993), Searle et al. (1998), Robertson and Collins (2002), and Borneman et al. (2015) present a range of scenarios involving basement deformation in the deeper areas of Kohistan and close to the Shyok Suture and Main Karakoram Thrust in northern Ladakh, whereby basement units are involved in a range of thrust, lateral strike-slip, and strong ductile deformations, with vertical movements of up to 10 km and lateral movements of 100–100s of km.

The Ladakh batholith mostly comprises crystalline plutonic rocks (and locally, meta-sedimentary rocks into which the batholith intrudes). It is difficult for granular crystalline materials to form thrust packages without the transformation of 3-D rock fabrics and the generation of foliation planes which can then act as slip-planes (Fig. 7.3). Almost undeformed crystalline granular granitoids are common amongst the less deformed parts of the batholith. These have no contiguous 3-D fabric although Fig. 7.3a contains incipient cleavage planes and evidence of plastic deformation. Figure 7.3b is a field photograph of granitoids beginning to form a cleavage, with individual crystals realigning, and forming a weak preferred orientation fabric. A strong penetrative cleavage fabric is present in Fig. 7.3c with d presenting strong gneissose and shear fabrics, highly-aligned minerals, some of which may be metamorphic or contain metamorphic growths (e.g. the hornblende crystals in Fig. 7.3d, and shear S–C fabrics. Well-developed fabrics facilitate significant slip-plane thrust deformation and the kinetic and friction-related heat energy generated by the thrusting process further accentuates and develops cleavage and fracture foliations.

Lacombe and Bellahsen (2016) present a comprehensive review of aspects of 'thick-skinned' tectonics involving basement focusing on Cenozoic aged orogens. They argue from a review of geophysical and geological data that there is increasing evidence for basement-shortening deformation within orogenic belts, but this can be complicated due to the heterogeneous and variable transmission of tectonic stress within a crystalline crust, with a wide range of inherited weaknesses and anisotropies.

Fig. 7.3 Transformation of undeformed granular textured granite to sheared gneiss. **a** Presents a coarse granular feldspar–quartz–biotite granite. Even this granite has incipient fractures which are beginning to stretch and align crystals in a bottom-left to top-right direction. **b** Also shows a feldspar rich coarse equigranular granite. Note incipient fracture/cleavage formation along weakly developed cleavage planes from top left to bottom right. **c** Presents well developed fracture-cleavage within a medium grained granite. **d** Presents sheared gneissic diorite with local S–C fabrics. Pencil is 15 cm long with a 2 cm silver tip. Photographs from the Phyang and Taroo valleys

The ambient pressure–temperature environment, crustal rheology, local boundary conditions, and structural-thermal inheritance parameters all add complexity to deformation styles and processes, and stress-transmission. These basement heterogeneities and structural inheritances can be a key control on the geometry, kinematics and mechanics of deeper-crustal deformation, and result in local variations in structural curvature, segmentation, along-strike variation, percentage of shortening, and sequence of deformation. Stress and shortening can be accommodated throughout large thicknesses/areas of crust and can be less obviously apparent compared with thin-skin-cover situations. Localised basement thrusting can require inherited structure, and remnant heat or heat generated through crustal underthrusting or locally generated tectonics. There may be significant differences or even a complete decoupling of structural styles between basement and cover.

Evidence presented throughout this study corroborates many of the conclusions of Lacombe and Bellahsen (2016). The southernmost High-Strain-Zone part of the batholith from Taroo–Phyang—lower Leh valley—lower Saboo–Shey–Thiksey—lower Stakmo–Stakna is strongly deformed by thrusting and local shearing and has a common structural style. It is in this zone where several flower-structure divergent thrust directional styles are present, local intense shearing-mylonitisation-partial melting is present (e.g. Stakna, Thiksey). This zone extends to its northernmost limit within the Leh valley, e.g. around the Shanti Stupa and the Sankar valley, which exposes several thrust-stacking-duplex and convergent thrust zones. This zone evidences the ability of the basement to shorten and transmit stress, with variable percentages of shortening and deformation along strike within the overall high-strain zone. The presence of areas with higher densities of shear-zones and an overall higher-strain ductile style juxtaposed to areas dominated by thrust faulting/lower strain areas with few thrusts, more normal styles of faulting, and relatively gently folded sedimentary sequences, suggests that different crustal levels are present within the study region. Each zone will bring with it pre-orogenic heterogeneities and inheritances, from sedimentary bedding and extensional-faults, to fractures, pluton–pluton contacts, igneous layering, and so forth. The complex structural style of the Ladakh batholith as compared for example with the adjacent Zanskar structures, and the consistent north–north-eastern thrust style of the molasse to the south of the batholith, indicates a strong element of deformational decoupling between cover and basement.

Analysis of thrust directions, coupled with the juxtaposition of zones of differing crustal depth, can be explained in terms of a fractured basement. The Ladakh batholith basement in the Leh region is acting like a series of discrete and separate crustal blocks which have variable lateral and horizontal motions relative to one another. The presence of relatively close crustal units which present opposite thrusting polarities can be explained by a model that allows crustal units to move independently. Figure 7.4 presents a model depicting separate crustal blocks, separated by faults, which can move independently, accommodating south- and north-directed thrusting in addition to other lateral and vertical stresses. This work proposes a model where: (1) the whole area was thrust southwards during Eocene–Oligocene times; (2) the crust fragmented splitting into discrete crustal blocks; (3) some crustal blocks thrust northwards during Mio-Pliocene times: those basement units that were particularly strongly coupled with the concurrently northward-thrusting cover sequences would have moved most during these times.

Figure 7.5 presents interpreted and uninterpreted satellite imagery of part of the Ladakh batholith close to, and east of Stakna, around Taktok Gompa. This crustal zone is highly fractured with locally exceptionally high densities of fractured crust. Lineaments trend from NNW to NNE and E–W to WNW–ESE. There are major structures that extend 100s of m and smaller structures at the 10s of m scale. Smaller structures have a clear link to larger structures, and are interpreted here as indicating significant lateral movement. Interpretations of relative motion, (i.e. left-lateral or right-lateral) are dependent on crustal depths and temperatures, and whether the structures formed in a brittle or ductile regime. Whilst much of the fault pattern is suggestive of a brittle stress regime, some structures have the S-shaped form asymptotic to bounding faults symptomatic of shear-ductile structures (e.g. top far right of Fig. 7.5). This extremely well exposed crust provides evidence for high-density fracturing of the basement in the region and demonstrates heterogeneous expressions of stress within a basement context. The crustal unit on the western side of Fig. 7.5 is characterised by north–south larger faults, and east–west smaller faults, strongly discordant to the east–west trending larger scale structures on the eastern side. A

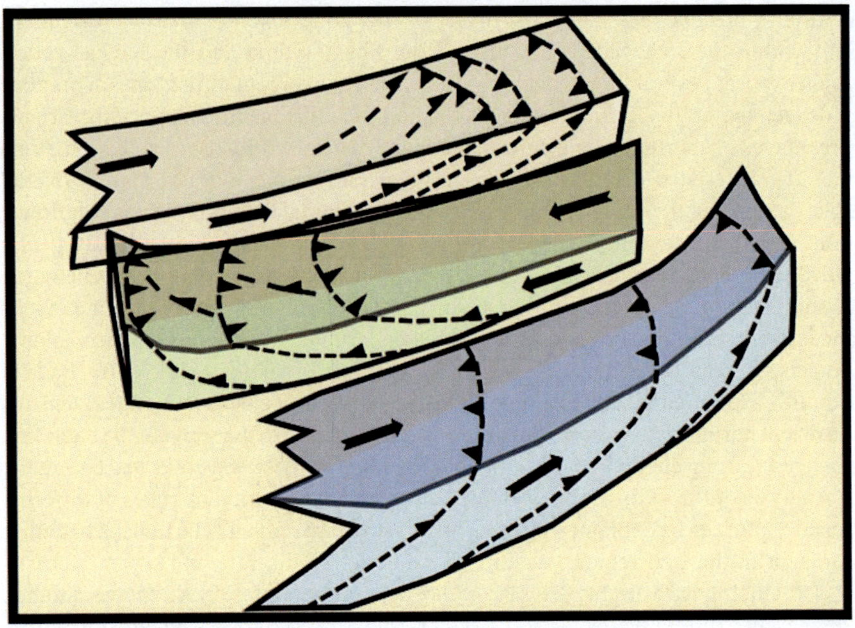

Fig. 7.4 Block model for the Ladakh basement geology. The basement is divided into a series of blocks divided by lateral fault systems. This may facilitate aspects of north and south directed thrust systems. The fractured nature of the basement allows for different levels and depths of crustal exposure

NE–SW river valley, whose trend is controlled by basement structures, separates the western and eastern units. The eastern crust itself is divided into several (>5) discrete blocks, separated by larger scale east–west trending faults. The sigmoidal character of some of the lineaments is a clear expression of relative lateral motion between various crustal units. Detailed analysis of the lineaments suggests that some blocks moved eastwards, whilst others moved westwards, and others north or southwards: all within a crustal area measuring c. 800 m by 400 m. There are close similarities between the generic style as suggested in the conceptual diagram of Fig. 7.4 and the actual crust as presented in Fig. 7.5. What is less evident in Fig. 7.5 are relative vertical motions between the discrete crustal units. However, if relative lateral motion is possible, so must vertical motion, and some/all the faults present in the image may have an element of vertical displacement.

Figure 7.6 presents a highly fractured piece of basement at outcrop scale to complement the larger scale satellite imagery of Fig. 7.5. The figure presents a range of micro-topographies with discrete units displaying both vertical lateral relative motion to one another. The largest fault on the right side of the image has a curvilinear form and clearly displaces fabrics either side of the fault. The curvilinear fault is a fault zone some 50 cm to > 1 m wide, characterised by a high density of faulting/fracturing adjacent to the main fault-plane, producing a localised lower

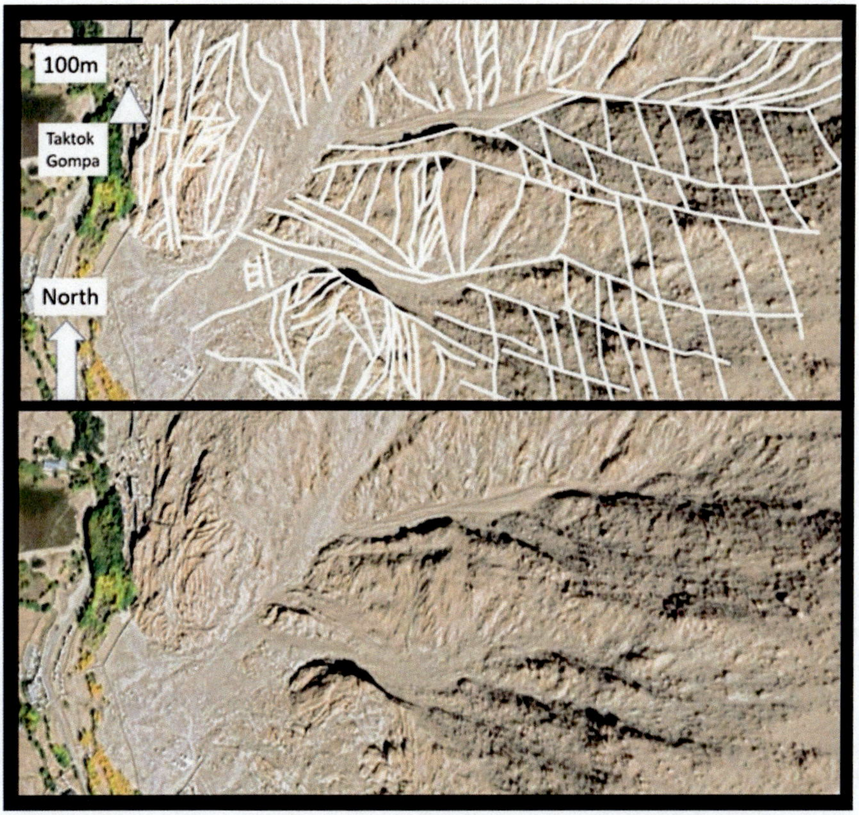

Fig. 7.5 Interpreted lineament map from the Taktok Gompa area of Ladakh (c. 30 km SE of Leh and 12 km east of Stakna). This well exposed section of Ladakh batholith basement exhibits a high density of fracturing and faulting. NNW–N–NNE to East-ESE fracture systems dominate the structure. The image evidences a complex structural history, with large fracture systems accommodating vertical, lateral, and trans-tensional/pressional deformation. Satellite imagery, Google Maps, accessed September 2022

topography 'rift' relative to crustal units left and right of the fault system. Rocks in the upper part of the image are at the highest local topographic/structural levels facilitated by faults oriented from top right to the bottom left of the image, and faults in the mid-upper part of the image, that trend from the upper left-hand side to mid-right of the image. The evidence here indicates that: (1) crustal fault blocks can move vertically and laterally within the Leh region; and (2) the basement can be highly fractured.

Developing and applying evidence and ideas as presented in Figs. 7.5 and 7.6 and accompanying narratives, Fig. 7.7 presents a fractured crustal block model for the Leh–Ladakh area. NE–SW and NW–SE trending faults break up the region into fault blocks that measure > 20 km NE–SW and 3–20 km NW–SE. The fault blocks present different crustal levels. The deepest levels exhibit sheared crust with locally

Fig. 7.6 Large outcrop scale Ladakh dioritic basement exhibits a dense network of fracturing and faulting as well as gneissic fabrics within the plutonic rock. Even at this outcrop scale the basement is divided into blocks separated by faults which facilitate both vertical and lateral movement, producing a micro-topography of relative basement highs and lows. Shanti-Stupa, Leh area

stronger gneissose and schistose fabrics and more limited amount of brittle thrust deformation. The medium depth crustal blocks are dominated by brittle–ductile thrust deformation to variable degrees, whilst the highest crustal blocks are dominated by low densities of brittle thrusting and faulting, the presence of larger outcrops of pre-batholith sedimentary rocks, mostly folded, with relicts of original horizontal bedding locally present, and high angle, mostly normal, but also reactivated reverse faults.

What emerges from this analysis is a well-exposed piece of fractured basement blocks impacted by several crustal tectonic stresses. At the larger scale the basement is sandwiched between the Indus Suture Zone and more southerly cover sequences deformed by both south and north directed thrust sequences, and the Shyok Fault System to the north. It is also underlain by the underthrusting northwards-directed Indian Plate, and a northern crustal Karakoram crustal 'buffer' with bounding thrusts and faults. At a regional to local scale the area is impacted by thrusting, extensional faulting producing the Indus Molasse basin, the relative rise and subsequent denudation of local mountains, and the development of the Indus and Shyok drainage systems.

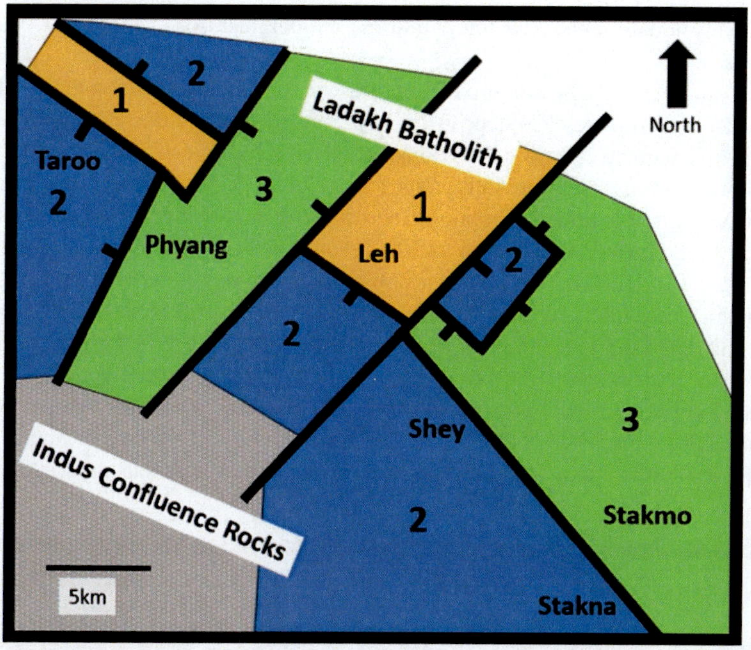

Fig. 7.7 Block model basement model for the Leh–Ladakh area. Thick black lines are faults with ticks on the downthrown side. Number 1–3 refer to relative crustal levels exposed within the batholith, with 1 denoting deepest crustal levels, and 3 shallowest crustal levels. The NE–SW and NW–SE faults allow for lateral and vertical accommodation of deformation within the Ladakh batholith basement

One interpretation of the geochronological data is that deformation is pinned between the intrusion age of the batholith and the age at which the crust reached the 150 °C isotherm level: or between 50–45 Ma and 30 Ma. This study suggests that deformation occurred over a more protracted period, probably from the Eocene to the Pliocene. The predominance of northward-directed thrusting within the Indus Molasse from 15 Ma is noted: this work suggests that this deformation has basement expressions. For this style of deformation to occur during Mio-Pliocene times the southern batholith must have been at temperatures that at least locally exceeded the threshold temperature for quartz–feldspar plastic deformation. This can be accommodated through: (1) high geothermal gradients and tectonism at low crustal levels; (2) moderate geothermal gradients and tectonism at deeper crustal levels and (3) local friction-induced heating facilitating thrusting. This model would default to option (2) as the most plausible model with deformation occurring at depths of c. 7–8 km. The overall model of Kirstein et al. (2009) may apply in general terms, but this study also suggests that the basement was uplifted, denuded, and buried variably with local block structures influencing these processes. Furthermore, the style of deformation present at Taroo, south Leh valley, Stakna, and other areas would require ambient (not merely locally enhanced temperatures) temperatures of at least 300 °C at c. 15–10 Ma.

The complete Ladakh batholith magmatic-sedimentary-deformation model (Table 7.2, Fig. 7.8) comprises: (1) deposition of pre-intrusion turbiditic probable-marine sedimentary sequences within the Ladakh arc during Jurassic–Cretaceous times; (2) intrusion of plutons and minor intrusions, locally from 70–50/45 Ma; (3) diachronous (50–30 Ma) variable uplift/denudation and surfacing of parts of the batholith, unconformably overlain by the Indus Molasse and eroded by deeply-incising river systems; (4) extensional faulting from 50 Ma accommodating Indus Molasse sediments to at least 5 km. These faults cut into the basement in the vicinity of the southern batholith. This constituted the beginning of basement fragmentation. Pre-batholithic turbidite extensional fault systems may have exerted at least local controls on the molasse basinal structures; (5) mid-crustal basement blocks thrust southwards, whilst deeper crustal blocks experience ductile shear stresses from Eocene–Miocene times. Local high deformation mylonite–cataclasite zones within thrust units causing in situ partial

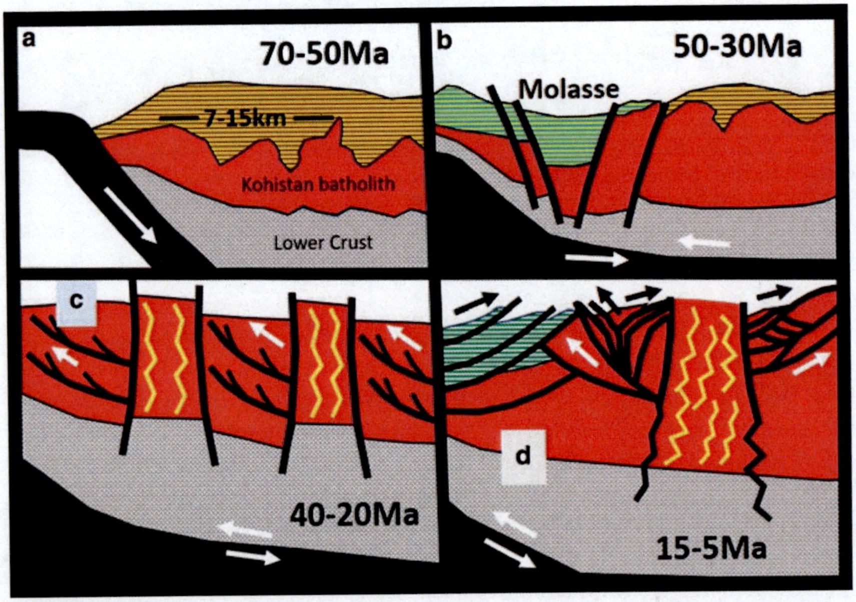

Fig. 7.8 Tectonic-Structural-Deformational Model for the Ladakh batholith in the Leh region. **a** The Kohistan batholith intrudes pre-batholithic arc marine turbidite sediments predominantly from Cretaceous–Eocene times. **b** Extensional rift deformation accommodates Indus Molasse and begins to dissect the batholith, whilst also eroding and unconformably depositing sediment on top of the batholith (Eocene–Oligocene); **c** batholith broken into a series of independent blocks with a range of lateral and vertical motion. Upthrown blocks preserve shear deformation, whilst shallower crustal levels affected by south-directed thrusting (Eocene Miocene); **d** north directed thrusting dominates impacting Molasse and basement with piggy-back, flower, pop-up and back-thrusts created within the domainal basement (Miocene–Pliocene)

melting. The basement is fractured and broken up into discrete blocks; (6) North-eastward thrusting of the Indus Molasse from the Miocene, with strong coupling in parts of the basement, moving some basement blocks northwards; (7) Uppermost crustal units folded, later local cleavage deformation produces fractures intruded by granitic veins and dykes syn-post deformation. Last deformation stage produces high-angle faults.

References

Ahmad T, Thakur VC, Islam R, Khanna PP, Mukherjee PK (1998) Geochemistry and geodynamic implications of magmatic rocks from the trans-Himalayan arc. Geochem J 32:383–404. https://doi.org/10.2343/geochemj.32.383

Borneman NL, Hodges KV, van Soest MC, Bohon W, Wartho J-A, Cronk SS, Ahmad T (2015) Age and structure of the Shyok suture in the Ladakh region of northwestern India: implications for slip on the Karakoram fault system. Tectonics 34:2011–2033. https://doi.org/10.1002/2015TC003933

Boyer SE, Elliot D (1982) Thrust systems. Am Asso Petrol Geol Bull 66(9):1196–1230

Burg J-P, Arbaret L, Chaudhry MN, Dawood H, Hussain S, Zeilinger G (2005) Shear strain localization from the upper mantle to the middle crust of the Kohistan Arc (Pakistan). In: Bruha D, Burlini L (eds) High-strain zones: structure and physical properties, vol 245. Geological Society of London Special Publication, pp 25–38

Coward MP, Jan MQ, Rex D, Tarney J, Thirlwall M, Windley BF (1982a) Geotectonic framework of the Himalaya of N Pakistan. J Geol Soc London 139:299–308

Coward MP, Jan MQ, Rex D, Tarney J, Thirlwall M, Windley BF (1982b) Structural evolution of a crustal section in the western Himalaya. Nature 295:22–24

Coward MP, Windley BF, Broughton RD, Luff IW, Petterson MG, Pudsey CJ, Rex DC, Khan MA (1986) Collision tectonics in NW Himalayas. In: Coward MP, Ries AC (eds) Collision tectonics, vol 19. Geological Society of London Special Publication, pp 203–219

Coward MP (1980) The Caledonian thrust and shear zones of NW Scotland. J Struct Geol 2:11–17

Coward MP (1982) Surge zones in the Moine Thrust zone of NW Scotland. J Struct Geol 4:247–256

Coward MP (1985) The thrust structures of southern Assynt, Moine Thrust Zone. Geol Mag 122:595–607

Holdsworth RE, Strachan RA, Alsop GI, Grant CJ, Wilson RW (2006) Thrust sequences and the significance of low-angle, out-of-sequence faults in the northernmost Moine Nappe and Moine Thrust Zone, NW Scotland. J Geol Soc London 163:801–814. https://doi.org/10.1144/0016-76492005-076

Honegger K, Dietrich V, Frank W, Gansser A, Thoni M, Trommsdorff V (1982) Magmatism and metamorphism in the Ladakh Himalayas (the Indus-Tsangpo Suture Zone). Earth Planet Sci Lett 60:253–292

Jadoon UK, Lin D, Baral U, Qasim M (2019) Early cretaceous to Eocene magmatic records in Ladakh arc: constraints from U–Pb ages of Deosai volcanics, northern Pakistan. Geol J 1–14. https://doi.org/10.1002/gj.3730

Kirstein LA, Foeken JPT, Van der Beek P, Stuart FM, Phillips RJ (2009) Cenozoic unroofing history of the Ladakh Batholith, western Himalaya. J Geol Soc Lond 166:667–678. https://doi.org/10.1144/0016-76492008-107

Kumar S, Singh B, Wu FY, Ji WQ (2012) Timing of Asia-India collision evident from U-Pb–Lu–Hf isotopes of zircons from granitoids and enclaves of Ladakh Batholiths, NW Himalaya. J Nepal Geol Soc 45:164–165

Kumar S (2020) Schedule of mafic to hybrid magma injections into crystallizing felsic magma chambers and resultant geometry of enclaves in granites: new field and petrographic observations From Ladakh Batholith, trans-Himalaya, India. Front Earth Sci 8:551097. https://doi.org/10.3389/feart.2020.551097

Lacombe O, Bellahsen N (2016) Thick-skinned tectonics and basement-involved fold-thrust belts: insights from selected Cenozoic orogens. Geol Mag 153:763–810. https://doi.org/10.1017/S0016756816000078

Ravikant V, Fu-Yuan W, Wei- Qiang J (2009) Zircon U-Pb and Hf isotopic constraints on petrogenesis of the Cretaceous-Tertiary granites in eastern Karakoram and Ladakh, India. Lithos 110:153–166. https://doi.org/10.1016/j.lithos.2008.12.013

Robertson AHF, Collins AS (2002) Shyok Suture Zone, N Pakistan: late mesozoic-tertiary evolution of a critical suture separating the oceanic Ladakh Arc from the Asian continental margin. J Asian Earth Sci 20:309–351

Rolland Y, Pecher A, Picard C (2000) Middle Cretaceous back-arc formation and arc evolution along the Asian margin: the Shyok suture zone in northern Ladakh (NW Himalaya). Tectonophysics 325:145–173. https://doi.org/10.1016/S0040-1951(00)00135-9

Rolland Y, Maheo G, Pecher A, Vila IM (2009) Syn-kinematic emplacement of the Pangong metamorphic and magmatic complex along the Karakorum Fault (N Ladakh). J Asian Earth Sci 34(1):10–25

Scharer U, Hamlet J, Allegre CJ (1984) The Transhimalaya (Gangdese) plutonism in the Ladakh region; a U-Pb and Rb-Sr study. Earth Planet Sci Lett 67:327–339

Searle MP, Cooper DJW, Rex AJ (1988) Collision tectonics of the Ladakh-Zanskar Himalaya. In: Shackleton RM, Dewey JF, Windley BF (eds) Tectonic evolution of the Himalayas and Tibet. R. Soc, London, pp 117–150

Searle MP, Weinberg RF, Dunlap WJ (1998) Transpressional tectonics along the Karakoram fault zone, northern Ladakh: constraints on Tibetan extrusion. In: Holdsworth RE, Strachan RA, Dewey JE (eds) Continental transpressional and transtensional tectonics, vol 135. Geological Society, London, Special Publications, pp 307–326

Searle MP, Khan MA, Fraser JE, Gough SJ, Jan MQ (1999) The tectonic evolution of the Kohistan-Karakoram collision belt along the Karkoram highway transect, North Pakistan. Tectonics 18(6):929–949

Searle MP, Hacker BR (2019) Structural and metamorphic evolution of the Karakoram and Pamir following India–Kohistan–Asia collision. In: Treloar PJ, Searle MP (eds) Himalayan tectonics: a modern synthesis, vol 483. Geological Society, London, Special Publications, pp 555–558. https://doi.org/10.1144/SP483

Treloar PJ, Rex DC, Guise PG, Searle MP, Windley BF, Petterson MG, Jan MQ, Luff IW (1989) K–Ar and Ar–Ar geochronology of the Himalayan collision in NW Pakistan: constraints on the timing of suturing, deformation, metamorphism and uplift. Tectonics 8:881–909

Treloar PJ, Brodie KH, Coward MP, Jan MQ, Khan MA, Knipe RJ, Rex DC, Williams MP (1990) The evolution of the Kamila Shear Zone, Kohistan, Pakistan. In: Salisbury MH, Fountain DM (eds) Exposed cross-sections of the continental crust. NATO ASI Series, C317. Kluwer Academic Publishers, Amsterdam, pp 175–214

Treloar PJ, Izatt ChN (1993) Tectonics of the Himalayan collision between the Indian Plate and the Afghan Block: a synthesis. In: Trelaoar PJ, Searle MP (eds) Himalayan tectonics, vol 74. Geological Society, London, Special Publications, pp 69–87

Treloar P, Petterson MG, Jan MQ, Sullivan MA (1996) A re-evaluation of the stratigraphy and evolution of the Kohistan arc sequences, Pakistan Himalaya: implications for magmatic and tectonic arc-building processes. J Geol Soc Lond 153:681–693

Treloar PJ, O'Brien PJ, Parrish RR, Khan MA (2003) Exhumation of early tertiary, coesite-bearing eclogites from the Pakistan Himalaya. J Geol Soc Lond 160:367–376

Treloar PJ, Vince KJ, Law RD (2007) Two phase exhumation of ultra high-pressure and medium-pressure Indian Plate rocks from the Pakistan Himalaya. In: Ries AC, Butler RWH, Graham